THE
SOFTWARE
SOCIETY

CULTURAL AND ECONOMIC IMPACT

William Meisel

Order this book online at www.trafford.com
or email orders@trafford.com

Most Trafford titles are also available at major online book retailers.

Cover design by SLM Designs

Printed in the United States of America.

ISBN: 978-1-4669-7411-1 (sc)
ISBN: 978-1-4669-7413-5 (hc)
ISBN: 978-1-4669-7412-8 (e)

Library of Congress Control Number: 2012924110

Trafford rev. 01/28/2013

 www.trafford.com

North America & international
toll-free: 1 888 232 4444 (USA & Canada)
phone: 250 383 6864 ✦ fax: 812 355 4082

For Susan, whose love and editing helped with this book,
and
For our children and their future

Table of Contents

Preface

As a technology industry analyst, I realized that there were important trends and issues that required a discussion deeper than my newsletter and blog could provide. Trends were making possible a tighter connection between *human intelligence* and *computer intelligence* with implications that deserved deeper examination. And, as software did more of what humans do, there were critical issues of the impact of that trend on jobs and the economy. Thus, this book.

Throughout a long career, I've been interested in the relationship between people and computers. An interest in how brains work and how computers could do some of the difficult things people do led to my teaching courses in Artificial Intelligence and computer pattern recognition as a professor in the Electrical Engineering and Computer Science Departments at the University of Southern California (USC) early in my career. I also wrote a technical book on how computers can recognize patterns. During the ten years I was manager of the computer science division of an aerospace firm, I applied this methodology to a variety of applications in defense, intelligence, and other areas.

Motivated to tighten the connection between people and computers, I founded and ran a company developing computer speech recognition. Afterwards, I became an independent analyst with a focus on voice and language interaction between people and computers. For more than a decade, I have written a paid-subscription industry newsletter (*Speech Strategy News*) and have edited two books with contributed chapters on the design of voice user interfaces. Currently, I consult and advise companies; organize conferences (most recently the Mobile Voice Conference, working with the non-profit Applied Voice Input Output Society, of which I am Executive Director); and I write a blog (www.Meisel-on-Mobile.com). These activities are a continuation of my interest in the connection between people and computers.

While this book was being written, the world was struggling with a recession, with some understanding of its initial cause, but little understanding of the slow pace of the recovery. I felt that software trends and automation in particular were having an impact on economics that wasn't fully appreciated. This book discusses the impact of accelerating automation and the increased power of software to do tasks that only people did in the past. I suggest an approach to faster recovery of the economy that recognizes the impact of accelerating automation.

In addition, some economists and analysts have expressed concern about whether new innovations in technology can compensate for the maturing of earlier innovations, leaving the US economy in particular in a slow-growth mode. This book suggests that the tightening of the human-computer connection is a development that could in fact spark continuing innovation and economic growth.

A balanced look at the impact of software on society must include other topics as well. One issue arises from the growing importance of consistency in our tighter use of technology. Since consumers and businesses will accept only a limited number of models for user interaction, it becomes important for a company to be a member of a community using one of those models. This trend has significant implications for business evolution. Further, problems with our patent system threaten this consistency, another issue that this book examines.

Still another issue of concern that arises from our increasing dependence on software is cybersecurity. A further concern is that the US education system in particular is not keeping up with this software evolution, both in terms of taking advantage of software in education and in terms of teaching computer literacy. With a goal of addressing the increasing role of software in shaping our society, I've tried to at least highlight the major trends and issues that this trend implies.

The admonition, "Make things as simple as possible, but not simpler," has been attributed to Einstein. Simplicity has virtues in its ability to help us reason about a subject and come to a useful conclusion—useful in that the consequences of actions taken based on that reasoning are what we expect. But *oversimplification* can leave out important factors and lead to bad decisions.

Oversimplification can lead to slogans rather than thought, and the application of principles whose only real virtue is their simplicity. In this book, I've suggested some "simple" ideas as guidelines for solutions to problems that are appearing. Yet, I've tried to avoid oversimplification. I hope I've achieved a balance.

In the text, I often specifically cite sources that I wish to attribute or quote. The bibliography at the end of the book includes those sources and others that have influenced me (in some cases sparking agreement and in others disagreement), as well as providing a resource for readers who want to dig deeper into the many topics I can only summarize.

Software innovations will also allow me to make this what I would call a "living book." After publication, the web site www.TheSoftwareSociety.com will contain a blog on specific trends, issues, and recommendations in the book, with the opportunity for readers to comment or make their own suggestions. I plan to participate in these blogs to react to comments, update my thoughts, and respond to current events related to the book.

Introduction

Software and People

Technology is a fundamental part of human history and a growing part of today's human experience. Our daily lives and interactions with one another are so interwoven with technology that one can't meaningfully discuss human society or economic development without incorporating the impact of technology.

Technology has always been an important part of society's evolution. Innovations in agriculture, transportation, and refrigeration improved our access to food and expanded our choices. Transportation improvements have changed our perception of distance and of how much of the world is available to us. The printing press made extensive retention and dissemination of knowledge possible. Telephones were an early expansion of human connections.

The more recent digital revolution is apparent everywhere. It seems as if every product uses a digital processor—from alarm clocks to microwave ovens. Personal computers are by now an old trend; the new trend is wireless devices, particularly mobile devices such as smartphones and tablet computers, which to a large degree let us take a personal computer with us wherever we go. The next big thing is smart TVs, bringing the interactive access to entertainment and information we enjoy on PCs and mobile devices to our living rooms. And the connection of these devices to the Internet gives us access to the seemingly unlimited computing power connected to the Internet and the immense data resources of the Web.

This digital trend will continue and accelerate. Every year, digital processors get more powerful and memory storage becomes more compact and less expensive. Access to the Internet, wired or wireless, will expand until it is available almost everywhere.

The increased digital processing power in our houses, our offices, our mobile devices, and accessible through the Web has made possible another trend—increasingly intuitive interaction with digital systems. Companies continue to innovate with visual displays, touch screens, speech interaction, and gesture recognition. In many cases, such as web searches and voice-driven "personal assistants," the processing is done in the network on powerful servers. As the complexity and variety of the things we do with digital systems explodes, these intuitive "user interfaces" become increasingly important to help us take advantage of the new capabilities.

Being connected to the Internet increasingly means being connected to our friends, family, and co-workers. Emailing, texting, tweeting, using social networks, as well as economical voice and video communications—all of them mean we can interact with others more flexibly and whenever we want. In developing countries, the growth of mobile phones has been phenomenal. People as well as digital devices are more connected.

We see some of these trends in our lives every day. We don't see as directly many of the areas where digital systems and software are impacting businesses and factories. Automation made possible by the increasing power of digital systems and the increasing power of software are making companies more productive, but, on the other hand, eliminating some jobs.

The changes created by digital systems are accelerating. The exponential growth in the processing power and affordability of digital systems is one part of the story. The other part, and one emphasized in this book, is the mutability of systems that are driven by software. Software is in fact "soft"—it can be changed without changing the hardware. When software is changed, the hardware does things it didn't do before.

We often update the software on our PCs or buy new software. We download apps in minutes to our smartphones. We can buy and begin reading a new book in minutes if we have an electronic book reader or tablet computer. The knowledge sources and services of the Web are updated constantly. This ability to change a hardware device by changing its software or connected sources of data, with no intrinsic additional cost, accelerates the impact of digital advances on society. In effect, the combined impact of hardware advances and software mutability are changing society more quickly than has ever been possible before. The remarkably fast growth of smartphones and tablet computers is a clear example of this trend.

Society is evolving through cultural evolution at a pace that makes genetic evolution almost irrelevant. We may feel that we have become accustomed to this pace. But there are times when an accelerating trend breaks through an invisible barrier and causes changes we don't expect—some good and some bad. It's a bit like a chain reaction—the impact can expand quickly, and, like a nuclear chain reaction, be destructive if not controlled. If controlled, it can generate a huge amount of useful energy. Software expansion has reached a point where its impact is central to the evolution of human society.

Two major themes are key in understanding the impact of accelerated technology change enabled by software. The *first major theme* is that we are getting increasingly connected to software, and that we can take advantage of that to couple computer intelligence and human intelligence. We can make computers and people more tightly tied together, with the tie more intuitive and always available to us because of mobile devices, thus increasing the ability of individuals beyond that of our intrinsic humanity. We can expect to always have a communication channel with other people available, both for immediate and delayed interaction. Children growing up today will expect a continual level of connection to computers and each other that was only science fiction a few decades ago.

This book will discuss how software is accelerating the growth of technology in general and personal technology in particular. Aspects of technology, such as recognizing speech, that seem almost human are actually done in much different ways than humans perform this task. Understanding the distinctions in how computers do complex tasks can help avoid the usual science fiction alarms about evil computers taking over the world. This book will make the distinction between human intelligence and computer intelligence, both in the way those types of intelligence operate and in what they can ultimately do. Exploring this difference is key in understanding how we may control the impact of technology's evolution on society to provide maximum benefits.

A key result of this advancing technology is a tighter connection between human intelligence and computer intelligence. There are many things computers do better than humans, e.g., they can accurately remember huge amounts of information. They can make that information quickly available to us on request. This tighter connection between humans and software is already deeply affecting our culture, and its acceleration can bring many benefits. It is beneficial to both individuals and to businesses that use this intensifying human-computer connection. This connection and its acceleration is a major theme of this book.

The *second major theme* of this book is that the accelerating capability of software and digital systems deeply affects the world economy. Technology has historically created the higher productivity that allows improvement overall in the economy. Through the higher productivity allowed by technological improvements, growth in national production could outpace population growth and allow, on average, improvement in conditions for individuals.

But technological change has always caused disruption in specific economic sectors, what the economist Joseph Schumpeter called "creative destruction." Two hundred years ago, some 70% of Americans worked on farms; today, less than 1% do because of farm machinery and other technology. Technology reduced the jobs available in agriculture, but slowly enough that workers adapted to the new jobs technology was creating, moving to cities where most of those new jobs were. Today, the disruptive effects of creative destruction are happening more quickly, and we must take action to adjust to that acceleration. The worldwide economic crisis occurring as this book is written reflects this issue; focusing on the direct triggers for the current crisis hides the more fundamental underlying hurdles to recovery.

Over-pricing of real estate allowed a boom that caused the economy to flourish more than it could sustain. But that trend hid a core issue in economics related to increasing automation. Automation can go beyond the point where there is time to adapt to it, I will argue; it can create a downward spiral. It is rational for every company to reduce jobs by automation if they can, so it is unlikely that any company will offer jobs it doesn't think it needs. Yet this philosophy overall in an economy can lead to high unemployment, overall lower sales in the economy, and the need for more automation to keep costs down. This is obviously a destructive cycle for an economy, but one in which we find ourselves today. Unemployment figures remain high, and slight improvement in the numbers hides the fact that some people are simply giving up looking for work. And statistics suggest that one of the groups hit hardest by current trends in automation is the critical middle class, with declining median incomes. Today, the benefits of increased productivity are being delivered unevenly, with most of the benefits going to the wealthy.

If jobs are eliminated overall, who will buy products and services? An economy requires customers. This is an aspect of economics that we are experiencing now, not just a future concern.

Humanity has managed to go through tough times—world wars, among other devastating periods of history—and recover. It is possible to manage this crisis and put long-term solutions into place without that level of pain, but recognition of the issue and quick action is required.

There are issues in any powerful technology advance, and software is no exception. As software grows in complexity, its behavior can be harder to anticipate and control. It can be used to abuse our privacy, for example. Software can disseminate misinformation as efficiently as it disseminates useful information. As we depend on software to control essential facilities such as traffic, communication, financial markets, and electrical systems, they become more vulnerable to malicious attacks, even attacks launched by foreign governments—attacks the source of which may be difficult to prove if they occur. This book will look at some of the issues this complexity raises and how they can be mitigated.

There are other areas affecting software's impact on society. One is in the area of patents, where "software patents" are creating increasingly difficult issues. Another is failure in education, particularly in the US—also part of the long-term economic problem. This book will outline problems and possible solutions in these areas.

Software trends increasingly drive economic trends. Some of the largest companies in the world are essentially software companies. The critical role of technology advances in improving economic conditions has been emphasized by economists such as David Romer in his groundbreaking 1990 research paper, "Endogenous Technological Change," and highlighted earlier by economists such as Schumpeter. The roles of improved productivity and creative destruction in moving society forward are increasingly recognized. Experts such as Daron Acemoglu and James Robinson (in their deeply researched 2012 book, *Why Nations Fail: The Origins of Power, Prosperity, and Poverty*) argue that elites fighting technological change to preserve their interests leads to the economic failure of nations. How will the accelerating change of technology impact politics and the very survival of individual nations? The tie between software and society isn't simply an academic subject.

We may hope that the economic issues simply go away, but wishing will not make it so. Jeremy Rifkin, in his 1995 book, *The End of Work*, raised the issue of technology eliminating jobs. He may have been early, but his concerns are valid. There are later expressions of concern over automation taking too many jobs, e.g., the 2011 *Race Against the Machine* by MIT researchers Erik Brynjolfsson and Andrew McAfee.

I have the advantage of writing this at a time that the impact of automation is evidenced in the slow recovery from the 2008 bursting of the real-estate bubble. As this is written, in the "recovery," US real Gross Domestic Product (GDP) per capita grew by 3.6% from its lowest point during the recession, but per capita employment fell by 1.8% from that point, beyond the 5.5% that was lost during the recession.

I believe the slow recovery is the result of companies investing in automation rather than rehiring people, and is structural, not temporary. The problem is worldwide, since this trend toward automation is international. Even in developing countries that use cheap labor as a driver of growth, automation has a key impact as expectations of their populace leads to better working conditions and better pay. I'm not predicting doom; there are economic factors and temporary fixes such as monetary policy that should give us time to address this issue, but only if we begin now. I also attempt to suggest specific solutions that I believe can work and are politically feasible rather than simply point out the problem. Among these solutions is an "automation tax," which, in a simple analogy, we might view as a payroll tax on computers—a tax that makes hiring a computer for a job more comparable to hiring a person.

The book is divided to emphasize the two main themes: The impact of accelerating software advances on our culture—our way of living—and on our economy, which allows us to live comfortably.

PART I

CULTURE

Software is pervasive in our modern culture, and its connection with people getting more intimate and pervasive. What do trends in software imply about what it means to be human? What are the barriers to full realization of the positive potential of the human-computer connection?

1

Major Trends in Software

Software in the form of computer programs is the machinery that drives digital systems. In that narrow sense, it is computer code. But a key is that it is indeed "soft." First, it is soft in that it can be easily changed and updated, much more easily than hardware. Second, it is soft in that it can be copied and shared without any significant cost of the sharing.

The characteristic of being able to share a good indefinitely without it running out applies to data and knowledge as well as computer code. Throughout this book, I will often use the term "software" in the broadest sense—code, data, and knowledge represented in that data—when discussing its impact.

The term "data" brings up the image of tables of numbers, but it is much more than that. The human body of knowledge is largely stored and passed on in the form of text, sound, and images; increasingly, that knowledge is preserved and organized in digital form and accessed by software programs. Knowledge and our ability to find it, create it, and share it is a fundamental trait distinguishing human society. Knowledge can grow with few limits if we preserve it and find ways to discover what we need to know quickly.

The focus of this book on software does not negate the impact of hardware advances. As I have noted, an increasing number of physical products incorporate digital processors and thus software. This expansion of "digital systems" is part of the trend of software becoming more pervasive in our lives. The expansion of systems driven by software will continue.

Why does software change the nature and speed of technology evolution and its impact on society? The following subsections highlight key trends accelerating that impact.

Exponential growth in computer power

The processing power available to software grows exponentially as we periodically upgrade our hardware, buy new hardware systems, or add servers to a network. Continuing improvement in digital processor speed and memory storage allows more complex software to operate quickly enough to be useful and to be affordable. Famously, Moore's Law says that digital processor and memory chips improve exponentially in complexity (doubling in the number of core processing elements—transistors—on the chips every 18 months). Exponential growth creates amazing progress—with Moore's Law rate corresponding to the number of transistors on a chip increasing by a factor of over 32,000 in 24 years. If that continued and was translated directly into computing speed, what takes 32,000 seconds (more than 9 hours) to compute today will only take one second in 24 years.

It appears that Moore's Law can continue for quite a while, with both improvements in basic processes and the use of multiple semiconductor layers on a chip (moving from a flat architecture to a 3D architecture). In a February 2012 talk where he predicted the continuation of this trend, Intel CEO Paul Otellini pointed out that Intel moved from a 32-nanometer process for chips in 2009 to a smaller 22-nanometer process for its Ivy Bridge chips in 2012, allowing more transistors per chip.

Nathan Myhrvold, a physicist by training and the former Chief Technology Officer at Microsoft, put this trend somewhat differently in a chapter of *Talking Back to the Machine; Computers and Human Aspiration* (edited by Peter J. Denning). He claimed that computing power has increased by a factor of one million in the last 25 years (speaking in 1999) and that it should increase by a factor of one million in the next twenty years; he felt this growth could continue for at least 40 years. That rate of growth means that in 30 years, computers of comparable price will be able to do in 30 seconds what it takes today a million years to do, he noted.

The trend may even be accelerating when translated into the support it can give software. Multiple microprocessor "cores" (processing units) are put on one chip, allowing more than one thing to be done at once on the chip—allowing more than one software program to be run simultaneously. For example, your smartphone might be finding news relevant to you while you dictate a message to be sent as text. Intel researchers were working on a 48-core processor for smartphones and tablets in 2012, targeting five to 10 years to market.

There are other methods on the horizon that could allow continuing growth in computing power if current approaches reach a limit. In October 2012, it was announced that David J. Wineland of the US National Institute of Standards and Technology would be awarded the Nobel Prize in Physics "for ground-breaking experimental methods that enable measuring and manipulation of individual quantum systems." Wineland's group has demonstrated computing operations based on the cited research that have the potential to accelerate computing power beyond today's technology. Wineland was quoted, "Most of us feel that even though that is a long way off before we can realize such a computer, many of us feel it will eventually happen."

Another aspect of processing power is the increasing use of network-based processing. For example, many smartphone applications use processing within the network. Companies such as Google compute search results on huge banks of computers in the network, not limited by the processing power of the device accessing these results. (More on this point in the next subsection.)

Obviously, computers will be able to use complex models and analyze quantities of data that are out of the question today. What we see today is impressive, but advances aren't stopping today.

The Internet and cloud-based computing

Internet connectivity, the World Wide Web, and search engines yield access to huge amounts of data—information that would not be feasible or easily updated on a local device. These huge data repositories go well beyond what any human could remember, of course, and in effect are an important adjunct to human intelligence. That intelligence is increasingly available to us wherever we are through mobile devices such as smartphones. Web search is often the first resource we turn to for activities such as finding a business or getting directions to an address.

The trend of fast access to Internet-based information is accelerating. In November 2012, Google launched its first installation of an ultra high-speed network to homes in Kansas City. The new technology claims speeds of up to 1 Gigabit per second. Currently, most Americans have Internet connections one-twentieth of that speed.

The Internet allows the power of software running in the cloud on powerful "servers" (specialized computers) to be available to connected devices. Thus, the resources for running the software can be much greater than what the device itself can support. And connectivity means that multiple servers in the cloud can support a single service. Further, this "cloud-based" computing power is available to businesses as a paid service, even to smaller companies, allowing the concentration of expertise, backup, and security in such services to be beyond what even some large companies want to invest. Netflix, the large provider of downloadable video, blamed an outage on Christmas Eve 2012 on the Amazon Web services it was using to serve customers. (Apparently, an Amazon engineer doing routine maintenance made an error that erased some data.) The largest Internet Service Provider (ISP, providing Internet access to servers for outside companies for a fee) is Level 3 Communications, which in 2012 provided services to over 2,700 corporations using over 100,000 miles of optical fiber.

Services delivered over the Internet (based in the cloud) are a major growth trend. Individuals look at such services, e.g., Google's web search and Facebook, as a natural part of the Web. The Web pages we visit are delivered by servers in the cloud.

And now companies big and small are increasingly looking at cloud-based services as an alternative to "premise-based" software, where software is purchased and loaded onto computers within the company. By using the Software as a Service ("SaaS"), companies avoid the costs of hardware and maintenance of the software. They can expand the volume of users without buying new hardware. What would be a capital expense becomes an operating cost, and is spread over time. The need to hire experts in that software to keep it running properly is eliminated, since the cloud service provider provides that function.

The cloud services provider gets the advantage of scale. The use of servers is spread over many customers, and the feedback from one customer leads to fixes to the software that benefit other customers before they even encounter the problem.

In terms of the quality of the software, the cloud-based approach has many advantages. Since the software in the cloud is fully under the control of the developer, without the burden of many different versions operating in many different environments on customers' premises, one version can be supported and updates can be very frequent, improving the customer experience and making it easier to add features.

A deliverer of cloud services can even compare versions of software before making them available to all customers. A common practice now is "A/B testing," where a company provides in the cloud two different versions of a service to different users (typically by random selection). The company can collect metrics on which version gets a better response (e.g., generates more sales), and adopt the better version across the service. Compare this to the old model, where a company releases a new version every year or two and ships it to customers on a CD, with no guarantee that they'll upgrade from a 10-year-old version, complete with all of its limitations. (This isn't a theoretical problem; many people, for example, are still using old versions of common programs such as Microsoft Office and old versions of operating systems such as Windows XP.)

In the 2012 book *How Google Tests Software*, James A. Whittaker, Jason Arbon, and Jeff Carollo comment that "Google has . . . benefited from being at the inflection point of software moving from massive client-side binaries with multi-year release cycles to cloud-based services that are released every few weeks, days, or hours." In an October 2012 shareholder letter posted on Microsoft's website, CEO Steve Ballmer indicated that one area of focus for Microsoft going forward was "building and running cloud services in ways that unleash incredible new experiences and opportunities for businesses and individuals."

The growth of cloud-based software has accelerated the rate of change of software beyond its intrinsic ability to evolve quickly. It is a fundamental trend that will continue, and it makes the acceleration of change driven by software even more pronounced.

Automatic software updates

Cloud services allow rapid changes in software while maintaining consistency, so that there is just one version to support, and companies that sell software that runs on client devices are looking to have this advantage for their software. Desktop software (and mobile apps) are moving in this direction, with automatic update ("auto-update"), updating the software without user intervention or with minimal permission requests becoming more prevalent. Major software developers are almost forcing updates, often with a warning that security concerns require the upgrade. Google takes it a step further with their Chrome browser, which automatically downloads and installs updates in the background (a feature you can't disable).

This may occasionally lead to some issues when the updates aren't invisible fixes to software problems. If a feature changes to be different than what the user is familiar with, it may cause resistance. There was one report that a significant feature was removed in a mobile phone software update that was characterized as a maintenance update by the vendor; the suspicion was that the feature was removed because of a patent suit by a competitor. If features start disappearing after one buys a product, it will cause a backlash, but, as I write this, the process hasn't been widespread or controversial.

The upside of automatic updates is that improvements can be added quickly, and the cost of supporting many versions is eliminated. This trend has so many advantages that it will certainly continue, despite occasional problems that arise. Again, this trend allows faster evolution of software.

Connected mobile devices

Smartphones and tablet computers with wireless connections to the Internet are making computer power available to individuals wherever they are, increasing our access to network-based information and computing power (and the software running it all). Smartphones deliver many of the same services we enjoy on our PC. A Verizon survey published in October 2012 found that 52% of all U.S. consumers surveyed say Internet service is their home's most important utility, and nearly 40% of Americans reported having an Internet-connectible device with them at all times, making them what the Verizon report called "borderless consumers." Over 800 million smartphones were sold in 2012. The trend toward "personal assistant" and "voice search" services make information more quickly and intuitively available. In addition, the expansion of search capabilities to the exploration of personal information on a device as well as in the cloud makes our own information more easily available to us; more and more of what we do is recorded digitally (e.g., our contact info and email) and searchable.

Wireless connections are becoming faster and more widely available. A number of wireless companies are moving to "4G" (Fourth Generation) networks such as LTE (Long-Term Evolution) to provide faster data rates, although shortage of available spectrum for supporting an increasing number of devices and increased use of the data channel for applications such as video may limit the speed of this growth.

A partial solution to the restraints on long-distance wireless technologies is short-distance wireless networks such as WiFi, which can supplement the long-distance wireless networks. Their reach is local for a wireless connection, but they use wired networks to get to the Internet. Smartphones can use a local WiFi connection if the owner enables it. WiFi and similar technologies can be found today in coffee shops, other retail establishments, offices, airports, many homes, and soon on many airplanes. It's safe to predict that this trend will grow.

The growth of "apps"

"Apps" is of course shorthand for "applications," and evolved as terminology for software downloaded to mobile devices. Mobile apps are obtained over the Internet and used on devices with an Internet connection.

One way of viewing apps is as an expansion of Web browsers. Instead of using a standard browser such as Internet Explorer, Safari, Chrome, or Firefox, the apps present an interface to the user that is tailored to specific tasks and thus more efficient. Some apps can run independently on the device without an Internet connection, but many are "distributed" applications, running partially on the device and partially in the network. One advantage of apps is that they can respond faster for some functions than a cloud-based service operating solely through a web browser.

There seem to be apps for every need. In September 2102, Apple was offering about 700,000 apps for iOS devices in its online App Store. The Google Play app store offered 675,000 applications for the Android operating system at the same time. Since an app can be downloaded in minutes, it essentially allows a user to customize their device, to add software that directly suits their needs. Apps also provide a new marketing channel, as "in-app" purchases generate sales.

According to a report in December 2012 from mobile app analytics company Flurry, US consumers spent an average 127 minutes on mobile apps per day, more time in apps than on the web and almost as much as they spend watching TV. They spent 43% of the time in mobile gaming and 26% of the time in social networking, according to the Flurry survey.

It may get more difficult to even know where software is running. As processors on mobile phones get more powerful and memory sizes increase, more of the software providing a service may be shifted to the device accessing those services, making the software even more responsive. The movement of some processing from server to client also reduces the cost of providing the services. The long-term result may be that cloud services run through apps will become increasingly powerful without proportional costs to the service provider, accelerating the trend further.

The Internet of things

The Internet lets people connect with people and with information or services on the web. Increasingly, devices are connected with other devices without a human in the loop. This has been called the "Internet of things" and the "Industrial Internet." (The last term may be somewhat narrower, referring to networks within specific industries.) In *Trillions: Thriving in the Emerging Information Ecology*, Peter Lucas, Joe Ballay, and Mickey McManus, refer to "pervasive computing" and "ubiquitous computing" as alternative names for the trend. They note that almost everything today has a microprocessor and devices that will be increasingly connected; they say we will live in "an environment of computation. Not computation that we use, but computation that we live in." The "trillions" in the title of their book refer to all the things being connected, far beyond the number of humans.

General Electric estimated in a November 2012 report that the Industrial Internet has the potential to add between $10 trillion and $15 trillion to global Gross Domestic Product by 2030. The report estimated that a 1% increase in efficiency created by the Industrial Internet could generate savings of $30 billion in aviation, $66 billion in power generation, and $63 billion in healthcare over 15 years. One example is sensors on rail cars that warn of impending breakdowns and prevent derailments; a major derailment can cost $40 million in damages.

The Internet of Things emphasizes devices talking to one another. The end result is to make the things humans use more efficient. A downside is that this trend will eliminate some jobs (reducing the number of railcar inspectors, for example).

More information

Easier access to digital data is driving the availability of more content in digital form. This adds value to software that can make the information more accessible and useful. The volume of information and opinions available as text and graphics over the Internet is impressive. The growth in short videos posted to the Web, for example (a fairly recent phenomenon), has been amazing; estimates put the number of videos on YouTube at over 100 million, despite YouTube only being founded in 2005.

Digital music downloads are the major way music is distributed today and causing waves within the music industry. Digital books are lowering the cost of books and are disrupting the printed books market and causing some bookstore closures.

The trend is another example of software driving change.

Software modules

Software functionality can essentially grow without effective limits because of its increasingly modular and layered nature, where each module can be improved separately by maintaining compatibility with other modules. The operating system (the OS, e.g., Microsoft Windows and Apple's Macintosh OS on personal computers or Apple's iOS, Google's Android, and Windows Mobile on mobile devices) runs programs that use the OS to save files, connect to the Internet, and do other basic functions without each software application having to include that functionality. The user interface contributed by the OS—e.g., pointing devices, icons, display, text entry, audio input/output, gesture recognition, and voice control—is highly complex software that each application doesn't have to duplicate.

Within application programs, "subroutines" break up software into manageable and reusable pieces. This modularity is one factor allowing exponential growth in software capabilities. Just a decade ago, there was often discussion of a "software crisis" caused by software getting so large it would almost be impossible to manage; there hasn't been much talk of the problems becoming unsolvable lately, although there are certainly issues of reliability and security.

Open-source software

Some software modules are "open-source," that is, the software is published in the form of the "source code," a readable version of the software in the form programmers write rather than the final version in binary code run by the computer processor (a form essentially unreadable). Open-source software comes with various license agreements, but many versions are available for free use in commercial products. A community of users typically contributes to suggesting features and pointing out issues. In the case of popular open-source software, it is widely used, tested, and understood, and provides a reliable software module that a developer can use with their application-specific software.

For example, about 60% of the servers on the Internet run Linux, an open-source operating system. Linux-based servers often run an entirely open-source software "stack," including the Apache web server (that supports the presentation of web pages in browsers, with about 65% market share), and MySQL database software (about 50% market share).

Google's mobile operating system, Android, is open-source. The number of smartphones using Android had grown to more than half of smartphones in use in late 2012, despite Apple's early lead (and still having about a fourth of the market to themselves).

There are many more such examples of the acceptance of open-source software. It is another trend that allows the rapid growth of reliable software.

Software continually does more

Software can implement inventions that are increasingly powerful in what they do, taking on difficult tasks such as finding the best driving routes or generating computer graphics. These methods are often based on mathematical methods requiring significant computing power. Continually increasing computing power makes it possible to run increasingly complex processes in an acceptable amount of time. Such software expands the tasks that computers can do for people.

Algorithms (mathematical methods implemented in software) that require significant processing of data to improve their performance (e.g., natural language interpretation, as in some mobile personal assistants) can be more effective because of the availability of increased computer processing power and the ability to use more data in *creating* (as opposed to *running*) the algorithms. We see the results of *using* algorithms (such as speech recognition) as a very quick response to an action we take (such as speaking a command), but the analysis to develop the speech recognition software handling a voice utterance probably took many days of computer processing. The author was involved in speech recognition research in the 1980s, and some analyses actually ran for months on a minicomputer that cost hundreds of thousands of dollars before any results were available; it was necessary to write the programs so that they could be restarted from where they stopped if the computer crashed or had to be stopped for periodic maintenance. Today, a mobile phone has more processing power and memory than that minicomputer, and a modern PC could run that analysis in hours.

The importance of this "off-line" processing power to software advances is often underestimated. Vlad Sejnoha, the Chief

Technology Officer at Nuance Communications (a firm providing speech recognition and natural language technology for dictation, mobile assistants, and other areas), estimated in a talk in 2012 that error rates in converting spoken words to text have been dropping 18% per year on comparable tasks, and that this can continue for the foreseeable future (perhaps we should call this "Vlad's Law"). In his presentation, Sejnoha summarized the source of the improvement: "Simply put, we're now using orders of magnitude more data, orders of magnitude more complex models to average and synthesize and capture those data, and orders of magnitude more computation to create those models during the training phase." Some of the more complex models are still challenging in terms of the computing power required to derive them, requiring "tens of thousands of compute cores in a parallel grid" in IBM's Blue Gene supercomputer, Sejnoha said. The resulting models, however, are efficient enough to run on today's commercial systems.

The message is that the increasing power of computer systems allows smarter software because larger "models" can be created with offline processing using more data. This trend will continue as part of the continuing advances in computer hardware.

Software advances allow new businesses to evolve quickly

Software can be the foundation of a business idea and allow that idea to reach customers quickly. One obvious case of a winning idea is web search technology, such as the Google search engine. While providing a free service to individuals, web search also generates huge advertising revenues, making the free service possible. The computing power in the network supporting web search emphasizes the role of software. Google said in a 2011 statement that its data centers continuously use enough electricity to power 200,000 homes; at the time, its centers around the world drew about 260 million watts, about a quarter of the output of a nuclear power plant.

Apple's innovations in smartphones and tablet computers are another example of how quickly products can be developed and build markets. Apple's innovations are largely in software; the company outsources manufacturing. The company's voice-interactive personal assistant Siri currently runs in the network, allowing its software to be updated frequently. The result has driven competition in smartphones and tablet computers and made Apple the most valuable company listed on the New York Stock Exchange as this is written.

Creativity gets easier access to markets

Digital systems allow an explosion of creative material, including video, audio, and books. The availability of these in digital form now makes them similar in ease of mechanical creation and duplication to software programs. Innovations such as digital books and word processing software, for example, make it practical to publish books by unknown authors or on narrow subjects that might not have survived the classical paper publication process. It's even relatively easy to start a web blog and have a chance of developing a following. The amount of such creative content accessible to consumers has grown exponentially as older content is converted to digital and the Internet allows almost immediate access. Impulse buying of books is more likely when immediate gratification is possible by downloading the book to an e-reader.

Personalization

Software is increasingly tied to individuals. It's likely that the reader has a "personal computer" that has been in fact personalized, with contacts, emails, a calendar, and other personal information and preferences. Our mobile phones have become personal assistants, using software in the phone and in the network. If our PC loses data or we lose our mobile phone, we increasingly feel that we have almost lost part of ourselves.

Some software also learns our preferences and interests automatically, through adaptation. A simple example is a word processing package that learns words (such as proper names) not in its dictionary originally and stops telling you they are misspellings.

Familiarity can breed comfort

As individuals who may have been resistant to using software are exposed to its advantages in environments other than learning to deal with a PC, such as mobile phones and Smart TVs, they are exposed to user interfaces that tend to be relatively consistent across devices in their basic operation, making it easier to adapt to the trend of increasing software content in all our devices. As "natural language" interfaces such as speech recognition and natural language processing improve, the devices become even more accessible, encouraging even the most technology-resistant to accept the help and resources that today's devices can offer.

Subsidizing hardware lowers cost and accelerates adoption

We see the price of hardware drop over time as part of the evolution of microprocessors and other digital components. Another factor accelerating the lowering of hardware costs is an increasing tendency of companies to subsidize hardware purchases. Mobile service providers have long subsidized mobile phones in exchange for service commitments. Beyond that continuing trend, some companies such as Google and Amazon are reportedly selling hardware near cost to obtain the income from other services provided by the devices, e.g., selling downloaded content or providing digital wallet services. These options will encourage adoption of mobile devices by consumers who are willing to try the devices at a lower cost.

Enabling faster change in technology

The big picture is that software allows faster changes in technology. Hardware is the part of technology that has a cost as a physical object. It changes more slowly, but the cost of hardware is dropping rapidly, particularly, as noted, with some companies subsidizing its cost. When that hardware contains some form of digital processing chip (e.g., the microprocessor in your personal computer), its functions can change very quickly, as simply as downloading an update to the software or a new application over the Internet. Access to ever-changing software in the cloud allows even faster evolution.

These trends making software an increasing part of society aren't independent trends. They are synergistic, and reinforce each other. These are the main factors that motivated this book and its observation that software evolution is accelerating, and is passing through a "tipping point" that changes its impact in fundamental ways, not all of them positive. Before delving into these complex societal impacts, it is useful to go a bit deeper into the nature of software and the hardware upon which it runs.

2

Software: More than Code

The word "software" probably brings up a mental image of lines of software code. That is indeed an embodiment of software, the instructions that drive a digital hardware system. In addition, software typically uses "data" and creates "data," which one might think of as a table of numbers or text stored in some form of digital memory system. But that simple characterization as code and data is somewhat like describing a human body as a collection of cells—it's true, but it certainly doesn't tell the whole story. This book will not explore in any depth how software is coded; the focus is software's impact on society. However, this section will attempt to give the reader unfamiliar with software some insights into how it is written and designed. In particular, this chapter emphasizes that delivering effective software goes well beyond simply writing software code.

The hardware

First, hardware is of course necessary for the software to do anything. The hardware is not limited to computers—PCs, "mainframes," servers, and the like. Most systems today, from cars to mobile phones to microwave ovens, depend on digital systems containing software to operate. The core hardware that makes software useful is processing chips—microprocessors—and memory chips, as well as other storage mediums such as hard disks or flash drives. The software is stored in binary code—0's and 1's—in memory chips or other digital storage and is interpreted by the microprocessor. Today, almost all systems store data and the program in the same memory system—the "von Neumann" architecture, first described in a 1945 report by John von Neumann.

Software as code

A software program is a series of instructions directed at the processor that tells the processor how to deal with the data in memory. As the program is "executed" by the processor, it may change some data in memory that represents intermediate results. The software may be interactive with a user, responding to user input from a keyboard, mouse, touch, speech, or other user interface, and delivering results through output devices such as a screen, speaker, or printer.

Software is generally written ("coded") in a "high-level" language, in which commands are written in text, with names that suggest their function. A programmer can then review the flow of the program with an easier understanding of what it is doing. The high-level language is converted by software into binary code before being "executed" by the processing unit.

As software has become more complex, two factors have helped keep it from becoming completely indecipherable and unreliable. One is the use of modularity, as discussed in the trends section. By breaking the software into manageable pieces, the programmer can more easily see what the program is doing, and can work on each module separately rather than one large program. When a module is useful in many programs, the programmer gets the benefit of reliable pre-existing modules without having to write them.

One important module is perhaps better called a "layer"—the operating system. Operating systems such as Windows on PCs or Android on smartphones do the heavy lifting in managing data in memory or hard drives, managing input and output, and many other functions that most programs can use without their being part of the program being written.

A second factor in reliability of software is the use of feedback from users. It's essentially impossible to test a program of any complexity for all possible conditions, so software almost universally has bugs. "Alpha" and "beta" versions of programs are usually issued in limited trials that uncover the most critical bugs. The final version of a software package will be tested through use by many more people in many more software and hardware environments, with resulting bug reports. Fortunately, bugs discovered by one user can be fixed before all users encounter them. This feedback process is critical to the quality of today's software.

In an interesting observation that could broaden the definition of programming, Brian Arthur in *The Nature of Technology* says, "Technology is a programming of phenomena to our purposes . . . The programming may not be obvious. And it need not be visible either, if we look at the technology from the outside." He is not talking of software code as the "programming," but he is looking at technology as an assemblage of components that is assembled according to some description in a "grammar" (analogous to a software language) suited to that technology. For example, a circuit diagram describes how to assemble components on a circuit board to achieve a certain goal. A blueprint may describe how to assemble an aircraft.

A programming language has a "grammar" of allowable ways to write commands, using the term in the same way your elementary English teacher might—saying "That's bad grammar" if you write "The apples is red." If a software program doesn't obey the grammar of the language, the programmer will get an error message when he or she tries to compile the program (convert it to binary code). Similarly, Arthur speaks of a technology domain: "A domain's grammar determines how its elements fit together and the conditions under which they fit together. It determines what 'works.' In this sense there are grammars of electronics, of hydraulics, and of genetic engineering."

One could stretch the definition of software to include all of technology using Arthur's logic. But this book emphasizes the rapid change possible by the pervasiveness of software in digitally based technology today.

Algorithms

Software code describes a series of steps that accomplish a goal. The goal may be relatively simple, such as computing an account balance in banking software. But, in many cases, the goal is much more complex.

Before one writes software code, one must have methods of achieving the goal of the software, often characterized as an "algorithm" when the method is sufficiently complex or mathematical. David Berlinski in *The Advent of the Algorithm: The Idea That Rules the World* defines an algorithm simply as "an effective procedure, a way of getting something done in a finite number of discrete steps." If an algorithm is defined as a "procedure," a computer isn't even needed; a written recipe is a form of algorithm with steps that a cook executes. A recipe even has a specific format: The list of ingredients essentially instructs us, "Make the following items available before you start."

But the simplicity of Berlinski's informal definition hides the importance of the concept now that computers can execute a huge number of "discrete steps" in a very small amount of time. In fact, he states: "The algorithm has come to occupy a central place in our imagination. It is the second great scientific idea of the West. There is no third." (The first great idea he refers to is the calculus, the math behind much modern engineering and physics.)

In practice, use of the term "algorithm" usually implies the use of mathematical techniques for precisely and compactly stated goals. The term is usually applied when the procedure is a particularly efficient way of accomplishing a task, perhaps motivating the use of the adjective "effective" in Berlinski's short definition of an algorithm. Often, an effective algorithm makes a task feasible that would otherwise be infeasible.

To get a feel for the subjective term "effective," let's look at a simple example. If one had two numbers a and b and wanted to compute aa + 2ab + bb, one could do so by the series of steps: multiply a by a, a by b by 2, and b by b, and then add the results of the three multiplications. This requires 4 multiplications and two additions. However, if one realized that aa + 2ab + bb = (a+b)(a+b) by simple algebra, one could use the algorithm: add a to b and multiply the result by itself, only one addition and one multiplication, an effective algorithm for computing aa + 2ab + bb. This simple example is typical in that the usual objective of algorithms is finding ways to do certain procedures efficiently, and it suggests accurately that an efficient algorithm requires insight and invention. If a computation such as our simple example had to be done billions of times, efficiency becomes important to obtaining a quick result.

More complex algorithms can do things like finding the shortest-distance route between two locations on a map, even making it the shortest-time route if current traffic information is available. It can recognize our speech and understand it well enough to do typical things we do with mobile phones, e.g., respond to "Give me directions to 123 Last Street," or "Text to Mary, 'I'm almost there.'" It can speak back to us in a synthetic voice generated by text rather than a recording, e.g., "Turn right at Green Street," using an algorithm that allows "speech synthesis" from text.

Algorithms can be implemented in software, but usually start out as a description in text and equations. An online review by a reader of the book *Introduction to Algorithms* cautioned readers: "Mastery of discrete math is a must; graph theory, programming, and combinatorics will also help." Books on standard computer algorithms may include chapters on esoteric subjects such as probabilistic analysis, sorting, data structures, graph algorithms, matrix methods, linear programming, and dynamic programming. The methods used for tasks such as finding the shortest route are based on mathematical methods that one can prove will in fact find the shortest route without trying all possible routes. Dynamic programming, invented by Richard Bellman in the 1950s, is still used today for tasks such as finding the shortest route between two points on a map. People such as Bellman, who wrote 35 technical books (and inspired me as a teacher in graduate school and was a member of my Ph.D. dissertation committee), are perhaps the equivalent in computer science to the pioneers in physics, but certainly don't get the same recognition.

Another broad category of algorithm is of particular importance to the theme of computers connecting with people (the subject of the next chapter). These are empirical methods, where data is analyzed by software to extract patterns, and those results used to perform a task such as speech recognition. (Such methods are sometimes called "statistical.")

One example of an empirical method is Optical Character Recognition (OCR), taking a "picture" of text (e.g., from a page scanned into the computer) and turning it into searchable and editable text. The software for doing so is created by processing many examples of text to extract the characteristics that allow distinguishing letters. The preliminary analysis of the examples of characters to derive the classification algorithm is a separate process from the embodiment of the results in software that does the character recognition on your PC or smartphone, a distinction I've highlighted previously. The analysis software takes a long time to derive parameters for a mathematical formulation that translates into fast software doing the actual OCR.

This type of process is called "pattern recognition." The methods used vary depending on the type of patterns being distinguished. I wrote a technical book called *Computer Oriented Approaches to Pattern Recognition* in 1972 and later applied the methods to many practical problems, including radar target identification, intelligence data analysis, air pollution episode prediction, economic forecasting, and speech recognition. Speech recognition, familiar today in call centers, medical practice, and mobile phones, uses a variation of pattern recognition techniques, analyzing huge quantities of speech and text in order to come up with a relatively compact model used to recognize speech while it is being spoken.

What computer pattern recognition can do

There are several factors that define where software can most easily automate pattern recognition. First, pattern recognition technology typically assumes the pattern is relatively consistent, that is, the factors that distinguish one pattern class from another are relatively unchanging over time. The way we write or type text is one example: If there weren't consistent ways to tell one letter from another, you would certainly be reading this much more slowly than you are. Software can perform Optical Character Recognition (OCR) because of this consistency.

This assumption of a fixed relationship is not absolute; things that change slowly can be followed through adaptation from continuing examples, making continuing adjustments to track the changes. A web search algorithm will soon learn the name of a new movie star, for example. On the other hand, it is hard to come up with a software algorithm that will predict the economy with any certainty because underlying factors that affect the economy change enough so that an example from 1950 might have little relevance to 2010.

Second, pattern recognition software requires a sufficient number of "labeled" examples from which to generalize. If it is doing OCR, it will have learned how to do so from many examples of letters in many fonts where the text is available both as an image and in digital text. This is one reason why it is difficult to automate something such as car repair or medical diagnosis; it's hard to find a large database of clearly defined symptoms versus diagnosis in either case. The increased use of Electronic Medical Records is a partial attempt to make more medical data available, so the dearth of usable data may eventually be reduced in healthcare.

However, the requirement for labeled examples is not an absolute. For example, there are methods researchers call "clustering" or "unsupervised learning" that help define similar cases without giving them a name. I've been told of some cases where consumers were clustered by software into groups that turned out to have similar buying characteristics based on a number of known facts about those people (e.g., the area where they lived). When a web site suggests other books you might like, the company is likely to be using technologies like cluster analysis to find groups of books that tend to be bought by the same individual over time.

Another factor that makes computer pattern recognition easier is if there is a clear set of quantifiable "features" that describe the pattern. For example, speech is typically first analyzed into a representation of its frequency content about 100 times a second for speech recognition algorithms. As long as a feature can be translated into a software subroutine to define it, it can be used; for example, subtle features such as "loops, whorls, and arches" that describe fingerprints can be translated into something that the computer can identify and locate. The requirement for quantifiable features is another reason jobs like auto service or medical diagnosis are difficult to automate; the symptoms are often described quite vaguely and in highly variable language even if they are documented. Service technicians and doctors learn to recognize patterns that they interpret as belonging to a specific class of problems and ask questions or do appropriate tests to narrow down the possibilities. There has been research for many decades in "expert systems" that attempted to emulate the step-by-step analysis of experts such as doctors, but this approach has produced few usable results.

One point that I hope the reader has gathered from this brief discussion is that the current methods computers use to recognize patterns are not an attempt to recognize patterns the same way the brain does. In a sense, since pattern recognition algorithms extrapolate from examples labeled by humans, they are learning to mimic the result, but not the methods, of human pattern recognition.

These empirical methods can be quite powerful, going beyond what people can do by analyzing data. Researchers at IBM, for example, determined that each of a baby's cries, from pain, to hunger, to exhaustion, sound different. IBM has patented a way to take the data from typical baby sounds, collected at different ages by monitoring brain, heart, and lung activity, to interpret how babies feel. An IBM web posting claims that, within five years, a mother will be able to translate her baby's cries in real time into meaningful phrases, via a baby monitor or smartphone.

Some researchers are trying to mimic the way the brain does things to a partial extent, however. At an event in October 2012 in China, Rick Rashid, the head of Microsoft Research, gave a presentation where he demonstrated live recognition of what he was saying and then the translation of that to Chinese using text-to-speech software that had been adapted to sound like his voice. Rashid summarized the underlying speech recognition technology as using a technique called Deep Neural Networks, which he said "is patterned after human brain behavior" and gave improved results over current methods. The methods are still statistical, however, since he said that the network models were trained with large amounts of speech.

Since our brains learn from examples, I suppose one could claim that computers and humans do pattern recognition similarly, but with different equipment. But I'd argue that the human brain seems to be doing things we don't completely understand yet, so I'm reluctant to compare our brains to hardware and software. A later section digs deeper into human versus computer intelligence.

3

Connecting People and Software

Software and the digital system on which it runs can act without human involvement. A thermostat turns heating or air conditioning units on or off depending on the target temperature. Interaction with the user (e.g., setting the temperature target) occurs relatively infrequently and is usually through a few simple buttons and switches. Some factory operations are fully automated, with devices rolling off an assembly line "manned" by machines doing repetitive procedures. In these cases, the interaction with humans is also typically in the set-up. There is minimal continuing interaction.

Until personal computers became popular, most people's interaction with computers was minimal. Early on, even programmers submitted decks of punched cards across a desk to be run on mainframe computers, later moving on to screen-based terminals that allowed more direct contact with the central computer.

Today, software and the systems on which it runs interact with people much more intensively and often. The word processing software on your personal computer probably has more features to interact with than you will ever use. Your mobile phone may try to emulate a human personal assistant, obeying your spoken commands and even bantering with you on occasion; that wireless assistant may even be addressable through your car's electronics. A computer game can provide intense human-computer interaction. An ATM will even give you cash.

In both the cases of minimal interaction and intensive interaction, there is some computer-human interaction method, in short, a "user interface." Even the simple on-off switch is a user interface, but complex software such as personal computer or mobile phone operating systems and web browsers allow the most complex user interaction.

The choice of the moniker "personal assistant" applied to some usability features in mobile devices (pioneered by Apple's Siri) suggests the growing sophistication of our relationship with digital systems. The intuitive nature of user interfaces will continue to improve over time, driven by improvement in the underlying technology, faster digital processors, improved connectivity with computer networks, and user feedback. There will be a continuing struggle between making software-based devices more usable while providing an increasing number of services and features that could compromise ease of use.

User interface design

One could argue that the source of Apple's remarkable success in innovation is its ability to design an effective user experience, from interaction with the applications on the device to the feel and look of the device itself. The skills involved in this process range widely, from an artistic sense of design to ergonomics—what fits human capabilities and limitations. The software must display results in a way easily absorbed by the user, with intuitive icons and menus that help us understand how to accomplish a task without referring to a manual. An underlying algorithm such as speech understanding, very complex software in itself, will ideally make a difficult task seem simple to a user.

Much of software user interface design qualifies as art. It requires an instinct and taste that comes in part from experience and study, but in many cases also requires an innate talent.

Systematic methods can make the process require less inspiration and guesswork. In a chapter entitled "Voice User Interface Design: From Art *and* Science to Art *with* Science" in a book *Speech in the User Interface* that I edited, Roberto Pieraccini and Phillip Hunter discuss how the process can be more scientific. For example, the software developer can test by using objective measures of performance to compare designs, or even by using mathematical techniques of adaptation ("machine learning") to automatically adjust a design to improve measured performance.

In any case, the importance of user interface design for software interacting with people is obvious. Systematic approaches to evaluating alternatives can reduce our dependence on intuition, but perhaps there is no substitute for inspiration in coming up with innovative alternatives to evaluate.

The evolution of human-computer interaction

Human evolution has always involved extending the core capabilities of our bodies and minds with external inventions. Our clothing extends the limitations of our bodies to withstand extremes of the environment, and, although we may change our clothing every day, it is almost part of us. Our automobiles in effect give us motorized wheels that go beyond what we can do with our legs alone. Our telephones extend the reach of our voice. We don't consider these extensions part of ourselves, but we certainly depend on them constantly. At work we use tools, ranging from hammers to copying machines, to go beyond what our bodies and minds alone can do. We transfer knowledge through books and media much more efficiently and to more people than face-to-face conversation would allow. Expanding our humanity through our inventions is nothing new.

Some connections with technology do become part of our brain. We drive a car, ride a bicycle, read, or type without thinking about all the detailed actions and processing needed to make this happen. They become part of our autonomous nervous system, embedded in our synaptic connections between neurons. We certainly didn't get these skills though evolution, yet we can use them even if we ignore them for years (e.g., riding a bike).

Digital systems can take the connection between humans and technology a step further, to a more conscious interaction, particularly when visual and language communication is used. The "coupling" between man and computer is increasingly tight and interactive. Somehow the user interface of a digital system has to use the characteristics of human senses and knowledge to make it easier for humans to understand how to accomplish a task using that system, whether a computer, mobile phone, or other device. The improvements in user interfaces have changed our expectations: We no longer expect to have to read a long manual before we use a software program, a device, or a service.

Users can interact with software through many modalities such as keyboards, a mouse, touch screens, voice, gestures (through a camera), and motion sensors, noting the result of their actions typically on a visual display or by voice feedback. There are many user paradigms for multi-purpose computing devices; the most familiar is the Graphical User Interface (GUI)—icons, menus, a pointing device, folders, windows, and layers of software engaged by clicking, double-clicking, right-clicking, finger motion, typing, or menu selection. The GUI is fundamental to most personal computers and mobile devices today. Our familiarity with GUIs is what has allowed us to ignore user manuals most of the time.

Ideally, the user interface should achieve its objective as simply as possible. But "as simply as possible" is getting more complicated—devices, software packages, and services are getting an increasing number of features and options. Simply having hundreds of songs on a portable device can make finding a particular one inconvenient using the GUI alone. Finding specific information on the Web is increasingly difficult as the amount of information and number of sources grows. A list of web sites that might contain the information you seek is less valuable than in the past, particularly if the search is conducted on a device with a small screen such as a mobile phone. The move toward "Direct-to-Content" solutions, using natural language understanding technology to try to understand the specific information or service one is seeking and deliver it directly, is an attempt to get one more quickly to the result one is seeking. The use of speech recognition to "just say what you want" (or respond to a more constraining prompt) is a relatively new paradigm, a "personal assistant" model of the user interface; this model is being extended to situations where one can't speak to "just type or say what you want"—"natural language interpretation." Classical search engines such as Google now allow typing or speaking, can handle some natural language requests, and deliver information directly (rather than a list of web sites) when possible, so the personal-assistant model could be considered an expansion of web search when the result is information from the Web. In some cases, the "personal-assistant" action goes beyond search, e.g., asking a mobile phone to send a text to a specific person with a specific message. Natural language interaction usually combines with the GUI for a unified experience, except when "hands-free" and "eyes-free" interaction—all voice—is desirable, e.g., when driving. As I write this, we are seeing the fastest evolution of user interfaces in history, in part due to the increasingly complex software supporting those connections and the rapidity with which it is changing.

The "personal assistant" model (whether one considers it a new paradigm or an extension of search) can connect with humans personally as the phrase suggests. It makes the human-computer connection more like a human-to-human connection (with the understanding that the machine need not be treated like a human, e.g., no need to say "please" or "sorry"). Apple has given Siri a personality in part by giving it a name, and in part through its answers to questions really not appropriate for a software entity (e.g., marriage proposals). Some would argue that we shouldn't think of a computer personality as a friend, but you could say the same thing about a pet if you wanted to be cynical. It's possible that, over time, some consumers will in fact choose hardware based in part on the personality of the main personal assistant user interface. I expect there will be specialized personal assistants that provide customer service for a company, and specialized personal assistants to entertain you interactively or give you advice on specific topic areas. Interactive cookbooks come to mind as an example, with the voice of the application perhaps being a popular cooking expert.

Algorithms play a large role in the evolution of user interfaces. Complex software has allowed a new generation of computer interfaces that go beyond the GUI, in some cases going around it and in others enhancing it. Two notable examples are speech recognition and speech synthesis (converting speech to text or text to speech), part of today's personal assistant model. Speech recognition can be used as an alternative to a keyboard to enter text, particularly useful on a small device such as a mobile phone. Text to speech synthesis similarly can read a text message aloud or issue directions to avoid the need to look at a screen. Text-to-speech technology has also been an asset to blind and visually impaired people, letting them hear text on a screen, part of what is sometimes called "assistive" technology, extending the human-computer connection to those who might have trouble with a pure GUI interface.

"Simple" text entry is aided by software on our digital devices. Spell checkers in our word processing program warn us of typos and grammar errors. "Predictive text" technology can help complete entries when we start typing; most of us have encountered predictive text when we begin typing a search term into a search box or a URL into a web browser and see suggestions of what we might be typing, either based on what others have meant when they started with the early text characters or based on our historical entries. Most mobile phones use "predictive text" when one is typing text messages. There are other text entry technologies; swipe technology allows entering text by moving one's finger from letter to letter on a touch keyboard without lifting a finger. Recent Nexus smartphones from Google even try to predict the next word, for example, suggesting "morning" and "night" as options following "Monday."

When one uses speech recognition, there is sometimes another layer required for effectiveness—natural language understanding, sometimes called "Natural Language Interpretation" (NLI). NLI can be applied to typed text or the text created by speech recognition. Search engines don't advertise it, but they are using a version of NLI to find what you want on the Web. To take a simple example, if you type a company name into a search engine, the top link is most likely to that company's web site, rather than to some web site that contains the words that form the company name. If you type "apple" in a search box, don't expect to get web sites talking about the fruit; the results displayed on the web page when I did so using Google were all about the company Apple. If you type in "apple pie," don't expect anything about the company to be in the results. If you type in "easy as apple pie," one of the top listings is a dictionary of idioms. A little thought about what has come to be an everyday experience in web search suggests the sophistication of the language processing that seems to return these results instantaneously.

More complex interactions using today's natural language technology may require a specific context to understand your intent. With a personal assistant application such as Apple's Siri, which attempts to understand a spoken request to a mobile phone or other digital device, the focus is what one usually does using the device.

Thus, it understands requests that translate to making a call, creating a text message or email to someone in your contact list, entering a web search request, navigating the streets to get to a destination, and entering or finding an item in your calendar/reminder application, among other tasks. Increasingly, such assistants can be tightly coupled to some Web sites, such as a restaurant review site to give you an assessment of how good your dinner might be.

According to a company blog in 2012, Google has deliberately avoided giving its version of personal assistant functionality a personality, preferring to simply call it an enhancement of its search engine, whether the search is by text or voice, and Microsoft seems to be similarly avoiding personalizing its Bing search engine. Samsung calls its voice personal assistant on its Galaxy III Android-based smartphones launched in 2012 "S-Voice," using technology provided by Vlingo, now part of Nuance Communications. Nuance has its own independent personal assistant app, *Dragon Mobile Assistant*, awakened by saying, "Hi, Dragon," so I guess they want a fire-breathing image (although they launched a version providing customer service named Nina in 2012). Perhaps vendors avoiding personalization simply want to avoid overly high user expectations.

The user interface between people and software enables our connection with machines, and, in doing so, can facilitate an immediate connection with other people (e.g., a phone call, video call, or chat) and allow communicating by delayed messaging (e.g., email, voice mail, text messaging, tweets, and social media postings). We can get access to another's thoughts less personally (e.g., electronic books, videos, blogs, and consumer evaluations of products and services on the web), expanding individual human abilities by letting us take advantage of others' intelligence to expand ours.

The human connection to machines has been tightening slowly over the years. The acceleration of this trend through software advances will have a dramatic effect not only on how we use machines, but on our emotional attitude toward them. Mobile devices, in particular, mean we can have a consistent connection to others and to computers wherever we are. Having the connection always available can make it almost part of us, like the auto when we are driving.

The trend of closer human-computer connections impacts us in fundamental ways that we are already seeing, evidenced in part by almost everyone having a mobile phone and by the incredible growth of smartphones and other portable devices such as tablet computers. Children growing up with these devices will ask their personal assistants questions as casually as we turn to 'web search for answers today. If you feel today's technology isn't up to this level of support today, keep watching.

Formal approaches to the design of interfaces between humans and computers have sometimes been called Human-Computer Interface (HCI) design. In the early days of computers, the design was task-oriented, since the machines largely had narrow functions such as entering data to specify a sales order or controlling the features of a copier. The methods used to create efficiency in such well-defined tasks aren't necessarily suitable to tasks that create an experience, e.g., game software. In *Technology as Experience*, John McCarthy (a computer scientist credited with having coined the term "artificial intelligence" or AI) and Peter Wright argue that any account of what is often called the user experience must take into consideration the emotional, intellectual, and sensual aspects of our interactions with technology. They suggest that the importance of how software is experienced is underestimated; design of user interfaces in their view is all about human experience.

One difficulty McCarthy and Wright cite in applying this insight is that the experience of the same user interface will be different for different users. They give the example of nurses who must enter patient treatment data in a computer, but view their main job as making patients more comfortable, so enter the data with the attitude that it is taking them away from the real job. On the other hand, a radiologist creating a report on an x-ray she is viewing realizes that another doctor will read the report to decide treatment, and the report is the demonstration of her expertise applied to the problem, the core of her job. For a radiologist, on the other hand, translating those observations into the structured forms of an Electronic Medical Record system may be viewed as "not my job." These attitudes—the experience associated with use of the technology—will affect the type of interface that can be utilized effectively.

One aspect of the human-machine experience that Apple's Steve Jobs perhaps recognized more clearly than most technology executives is the importance of beauty in design. It is hard to argue that we aren't moved when we see elegance, taste, or beauty in people or architecture, but it took some examples in the marketplace to show that beauty in design and materials was important with digital devices as well. Beauty can apply directly to software interfaces as well, for example, in use of color and icon design for static designs, motion in animated actions, and appropriate sounds to help call attention to an alert or action.

Elegance can lie more abstractly in the employment of good analogies that convey the use of software, e.g., the use of "folders" to organize our files and applications on PCs and "drag and drop" to move files or folders. These analogies are so familiar at this point that we no longer think about the association with real-world objects. A young child brought up with computers some day may say, "Daddy, is that folder in your filing cabinet like the folder on my computer?" To the child, a folder is likely to be something associated with a digital device rather a physical file cabinet.

Analogies to human experience outside the computer realm are a powerful way of making software-driven devices more useful. A pervasive example in graphical interfaces is pointing, whether we point with a mouse, finger, or stylus.

The interface isn't always based on something we know. We may be motivated to learn a new skill until it becomes part of our knowledge. The prime example of this is typing at a keyboard. We certainly aren't born with knowing how to touch-type, but it is a valuable enough skill that almost everyone today who uses a personal computer has developed it to some degree. We think about the words we want on the screen, and our fingers magically make them appear.

Clicking, double-clicking, or "right-clicking" with a mouse is now intuitive to personal computer users, although there is no clear analogy in "real life." (We don't double-click on a kitchen drawer to open it, at least not at this writing.) However, there are only so many skills of this sort that we are willing to invest in developing. Thus, the evolution of user interfaces has been in the direction of using knowledge and skills we already have.

The GUI is one of the most successful interfaces between man and computer. It depends largely on our basic intuition gained from intrinsic and learned experience. We see it in the way we interact with most devices today. The original personal computer model used what has been called the WIMP model (Windows, Icons, Menus, and a Pointing device). The original pointing device was a mouse, and that model has been extended to touch screens where our finger is a pointing device. The original idea keeps getting extended, building upon our familiarity with the basic model.

The basic model works by analogy. When we open a PC "window," we are not trying to get a breath of fresh air. We are looking through a "window" into a certain folder, document, web site, or application. When we "point" with a mouse, it is like pointing with a finger, but we move the mouse over a mouse pad to move the pointer, with the cursor or arrow showing us where the mouse is "pointing," but we think of it as if we are actually pointing to something with a finger. The connection in our mind between moving a mouse and the motion of the pointer on a screen is one of those things the synapses in our brain have learned so that we can use that connection without thinking.

An icon is a small picture that suggests what it means visually, appealing to what we know. We recognize the icon launching the Microsoft Word application because it is a stylized "W" and iTunes because it shows a musical note. The operating system represents document icons differently than application icons, so that we have a visual indication of what the item is.

Analogy in such user interfaces helps, but it only goes so far. We simply learn some things because they are consistently used by GUI-based operating systems and applications, such as clicking on a menu to see options drop down that we can select.

The classical Graphical User Interface has limitations. We are seeing, in my opinion, the limitations of the WIMP approach in personal computers and particularly in smaller devices today as the number of options overwhelm the paradigm. There are so many features in a typical application software package today that it can be a significant effort to find a seldom-used feature or even to discover if it is there. Onc could get away without a user manual in the early WIMP days; today, the "help" function gets all too much use.

Language in the user interface

The most sophisticated way humans obtain and transmit information is through language, and it is of increasing importance in man-machine communication as the technology for computer understanding of speech and text evolves to the point of high utility. Language is one of the skills that we invest a lot of time in learning and expanding over time, and it is to our advantage to use this fundamentally important means of communication if software is to reach its potential as a human aid.

We first learn spoken language and then invest a lot of time learning to communicate the spoken language in text form. The two aren't the same. They don't even use the same senses when perceived—speech uses our ears, and text our eyes. Usually, we can read or scan text much faster than we could listen to the same material and understand it, and it's easier to re-read than to re-listen. The two means of communication are fundamentally different. But our brains maintain a consistent connection between the words and concepts, whichever format the words are in. Language might be considered the most distinguishing and fundamental feature of humanity; it allows us to preserve knowledge and learn from one another and from the past.

But the essential point is that language allows deep and subtle communication of information and ideas. Since we use text extensively with computers and digital devices, it might seem that interaction via language is a well-developed man-machine interaction, but this view is misleading. The most common use of text on digital systems is to enter data that we interpret ourselves (e.g., a task list) or that we intend for other humans (an email or text message). Similarly, telephones were designed (originally) for people to talk with people—they just transmitted speech as electrical signals and reconstituted it as audio.

Human communication with computers is another matter. A conference on "artificial intelligence" (AI) was first held in 1956 discussing this subject, and most would argue that we have made very limited progress in that field to date if the definition of AI is a computer imitating a human so well we can't tell the difference (a test that the mathematician Alan Turing proposed).

Consider software that can recognize speech. Researchers tried for decades to get computers to work with speech. Speech recognition technology got better as methodology and databases available to build recognition software improved, and as more computer power to run the final result became available. Early solutions often worked with very limited vocabularies, such as the digits for number entry or a few commands for hands-free operation of software or equipment. As computers became more powerful, dictation with large vocabularies, "speech-to-text" software, became available. Early speech-to-text software worked best when the context was limited, e.g., a doctor dictating a medical report. For example, radiologists explaining what they saw on a chest x-ray would tend to use similar ways of expressing their interpretation (and wouldn't mention bones in the foot or try to order a pizza). But, over time, the dictation software got very good at general content as well, and software for transcription is purchased by many people today. In mobile phones, speech-to-text software is available to let us dictate emails or text messages without the distraction of a keyboard.

But transcription is not communicating with the computer. Communication requires understanding of intent, and speech understanding is a much more difficult task than speech recognition (speech-to-text) alone. Early versions of task-oriented speech recognition used limited context, e.g., voice dialing by spoken name in a mobile phone. The "understanding" by the phone was achieved in part by one having to be at a point in the phone's operation where one was entering a phone number, and the name spoken had to be in the phone's contact list.

Today, speech understanding has come a long way, although there is much room for improvement. The natural language processing that is the key to understanding is much less mature than the speech-to-text functionality.

The most complex speech understanding, e.g., by personal assistant applications such as Apple's Siri, Nuance Communications' Dragon Mobile Assistant, or advanced voice search such as Google's search engine or Microsoft Bing, is typically done within the network, where significant processing power is available and where the information sought by the spoken inquiry usually resides. Speech understanding technology has clearly broken through a barrier to utility, judging from the relative popularity of voice search and of personal assistants on mobile phones, enough so that Apple for a period of time featured Siri on expensive national TV advertisements. In general, "personal assistant" software accepts input in natural language and produces the desired result that is implicit in that input as best possible.

Another factor that will increase performance of natural language systems is personalization; the methodology can tune itself to your particular interests, location, and more. The methodology that allows this is often called "machine learning." Machine learning lets software adapt its behavior based on your behavior specifically or on the behavior of all users. It's sufficiently important to the future of user interfaces that Microsoft CEO Steve Ballmer mentioned it in his 2012 Letter to Shareholders.

As the amount of data available on the Web, on our personal devices, and in business databases increases, however, there is another level of software capability required. We need methods for "knowledge representation," describing what the answer is to a question and not just the fact that the answer is somewhere on a web site or in a specific document. We'd just like the *answer* to an inquiry, rather than a list of options. Again, software for such tasks is advancing as well. An example is IBM's Watson technology, which famously defeated two expert contestants on the *Jeopardy!* TV show, and is available today to enterprises to extract information from the excess of today's sources (a problem sometimes described as "big data"). Using the IBM Watson technology, IBM's Cognos performance management and business intelligence software can take vast quantities and different types of data, and transform them into "actionable insights," according to company announcements. Classical web search has in the past required us to painfully review multiple sources of information in multiple steps to find the specific information we want. Increasingly, personal assistants and search engines attempt to return the answer more directly. As a simple example, a request for weather will most likely report the local weather rather than just offering the option of going to a weather web site.

In a 2013 book *Mobile Speech and Advanced Natural Language Solutions* edited by Amy Neustein and Judith Markowitz, I argue in a contributed chapter that we can make search and personal assistants more "intelligent" by formal standards for representing content in sources such as web sites or web pages. Today, the answers we get when searching or making a request to a personal assistant application depend on the search application deciding what sources and specific content in that source is relevant. However, the developers of that source of information are best qualified to specify what they are providing in a standardized form, providing data that search applications could use in providing more direct answers.

Interaction by language also opens the option of "conversation," interacting with a search or assistant application to clarify a request. For example, Nuance Communications's Dragon Mobile Assistant can carry on the following conversation, according to company literature:

- User: "I need to create an appointment"
- Dragon: "When should it start?"
- User: "7 PM tomorrow"
- Dragon: "Here is your appointment, would you like to save, edit, or cancel it?"
- User: "Change the subject to "Dinner with John"
- Dragon: "Here are the best matches for John, please select." (If the user's contact list has more than one "John.")

This example illustrates how "personal assistant" applications can evolve to provide increased flexibility. In my opinion, the personal assistant model will dominate search in the future. One could extend the definition of "search" or "voice search" to include personal assistant functions, but this book will use "personal assistant" or "assistant" when the expanded capability is available. I emphasize, however, that the personal assistant should ideally provide the option of either text or speech interaction; there are times where noise, politeness, or privacy concerns make speech a poor option.

The personal assistant model seems to be gaining traction in China. At the end of 2012, the leading Chinese search engine, Baidu, launched the Baidu Voice Assistant for Android. The applications allow asking for songs, weather, and other features typical for personal assistant applications. It will even tell you a joke if asked. Another Chinese search engine, Sogou, released a smartphone app called the Sogou Voice Assistant; the company is investing heavily in speech recognition, with a research center with nearly 100 people tasked to create Siri-like software. iFlyTek, in which China Mobile took a 15% stake, offers "iFlyTek YuDian," a personal assistant application available to anyone with an Internet-enabled mobile phone. YuDian uses iFlyTek cloud-based speech recognition and provides access to entertainment, travel, and ticketing information. iFlytek is part of the Speech Industry Alliance of China (SIAC), created to advance Chinese-language speech recognition. Other members of SIAC include Lenovo, Huawei, China Mobile, China Unicom, and China Telecom.

Language provides a close coupling to our brains. It allows deep interaction with software. It allows us to use the incredibly large memory capacity and brute-force processing power of computers to extend our human capabilities beyond anything we can do without that close coupling.

A brain-computer interface!?

Some experts have predicted another level of coupling between people and software. They believe that we will some day have direct coupling to the brain. This goes beyond a skull cap that listens to brain waves, since that is very limited in the subtlety of communication possible (a bit like listening to the murmuring of a crowd to elicit its message). The ultimate man-machine communication, some futurists predict, is an implant in the brain that, like a wire that connects your computer to the Internet, connects you to the computer. For example, one of the founders of Google, Larry Page, was quoted in Steven Levy's book *In the Plex* as believing something like this will happen. Levy asked Page and co-founder Sergey Brin in 2004 what they saw as the future of Google search. Page replied, according to Levy, "It will be included in people's brains. When you think about something and don't really know much about it, you will automatically get information . . . Eventually you'll have the implant, where if you think about a fact, it will just tell you the answer."

An Intel-commissioned white paper from consulting firm Booz Allen Hamilton in 2012 on the future of mobile technology concluded that connected devices will inevitably interface with the human brain directly: "As convergence continues across device types, functions, and capabilities, the melding of mobile technologies directly into the human body becomes the logical next step." The June 2, 2012, issue of the *Wall Street Journal* had a feature article, "Bionic Brains and Beyond: High-tech implants will soon be commonplace under our skin and inside our skulls, making us stronger and smarter," by Daniel H. Wilson, a science-fiction writer. A recent book by Wilson is a story of those devices enhancing humans to such a point that it creates a conflict between the "have" and "have-nots."

The difficulty with the implant idea is that language enters our brain through our senses. It takes a while for a human to understand spoken language, and even longer to absorb language as text. What is the implant doing? Sending digital signals to a bank of neurons? If so, it's another way we have to understand language, since that is the way we process complex information. The implant would have to occupy a lot of neurons to get the level of detail that our auditory and visual nerves provide. And we'd have to learn to use those externally generated signals. Perhaps it's like typing, and it might be worth it. But the likelihood is that the implant would have to be inserted in childhood (perhaps even as an infant) to allow time for our brains to interpret its signals. I suggest reading *On Intelligence* by Jeff Hawkins and Sandra Blakeslee to get a feel for the complex multi-level process by which the brain learns to understand; one doesn't just put electrodes in the brain and expect to communicate!

And, how would we connect to the implant? Wires through our skull or a wireless transmitter (and battery of some sort) inside our head? The experimental subject for testing this technology would have to be highly motivated, perhaps by having his normal auditory and visual senses impaired. Perhaps in such a case, the technology would be worthwhile and successful, since the subject would be highly motivated. But this is obviously not the general case.

Such procedures would face tremendous practical hurdles. After all, inserting even a few wires into the human heart to treat heart conditions has proved hazardous; defective wires made by St. Jude Medical to connect heart-shocking defibrillators caused at least 20 patient deaths due to short-circuiting, according to an article published in a cardiology journal. Deliberately sending electrical signals to the brain could have risks of causing seizures and permanent brain damage, along with other unpredictable side effects. The failing defibrillator wires have led to issues of whether they should be removed surgically before they fail, a procedure with its own risks. And removing a brain implant is likely to be more dangerous than inserting it.

The practical hurdles of implants aside, it seems as if such an implant would accomplish nothing for a person with normal sensory abilities and language skills. Bluetooth wireless headset microphones couple to our auditory and vocalizing system quite effectively without surgery. Text on mobile devices can be entered and read quite efficiently (when one isn't driving). The direct connection to our brain today is through our existing senses.

As this is written, Google has demonstrated a "wearable computing" device that is worn like glasses. Information is displayed on a small screen to the side of one eye and communication with the device is by voice. Press reports indicate that Google will make early versions of the device available at a high price, almost as a high-fashion accessory initially, but it could evolve to a high-volume product. The advantage of the device is that it is always there—one doesn't even have to take a device out of one's pocket. It becomes part of us in a more obvious way.

While Google's eyeglasses are experimental, Motorola Solutions offers a commercial product, the HC1 headset computer, designed for use in industrial solutions where hands are occupied and mobility is important. The HC1 was announced in 2012 as the first in a new class of Motorola hands-free enterprise mobile computers that use speech recognition, head gestures, and video streaming to navigate applications that access and present business documents and schematics. The HC1 uses optical micro-display technology, providing the user with a view equivalent to a 15-inch laptop-size screen because of its resolution and placement near one eye. The speech recognition and natural language software supports six languages for application control, with dual bi-directional noise-cancelling microphones and a near-ear loudspeaker. The headset's accelerometer and digital compass deliver gesture control and direction and position orientation for navigating through applications. You can buy one today if you have the money and desire. And, like all technology of this sort, there will be improvements and price reductions over time.

A brain implant that used visual or voice simulation would add little, if any, value to those devices, even if one believes that one could learn to use the implant. The implication that an implant would instantly convey information to the brain defies close inspection. We understand with language and the understanding must be delivered in that form somehow and integrated with knowledge we already know. Information in the brain is stored in a highly distributed fashion, and it requires time to change the chemical bonds that constitute long-term memory.

Presumably the proponents of a direct brain interface are assuming some sort of more "efficient" direct digital connection than mimicking our senses. But how would that work? A direct digital input to the brain would be a series of binary numbers (probably with those numbers converted into a series of analog pulses simulating neural firings at different rates). How would we interpret that input? It is one thing to use a neural implant to stimulate specific sections of the brain to prevent a seizure or other targeted purposes, or to interpret external "brain waves" that identify which part of the brain is active, and quite another to use the technology for communication of complex ideas.

Input to our brains is one thing. What about output to the computer? How do we make an inquiry that the computer could understand in this mysterious binary language? We would have to learn to "talk" this language as well as understand it if the connection was to be of any practical use.

Other than these hopefully obvious issues, the idea that it is inevitable that a successful technology will somehow be tied to our bodies as a general practice doesn't pass the test of history. Wheels and motors are certainly key technology developments that have been around for a long time; they could be thought of as extending the function of our legs when we use them in automobiles, bicycles, or skates. Yet we prefer those wheels and motors to be external to our body, and not a permanent attachment. Why not just surgically attach motors and wheels to our feet and attach the controllers to our nerves? The idea of a direct brain connection to digital systems as a general practice is equally unlikely in this author's opinion.

I addressed this direct brain connection issue in some detail because I don't wish to return to it again or have the reader assume I'm ignoring the projections of some very bright people (despite my disagreeing with them on this point). Complex human-machine interaction using language through our existing senses is already at a state where it is valuable, and it will continue to get better. I focus on this proven and rapidly improving approach.

Human Intelligence versus Computer Intelligence

A theme in some science fiction books and movies is the computer that gains human-like intelligence and threatens mankind (or at least some of its members). HAL (Heuristically programmed ALgorithmic computer) in the movie *2001: A Space Odyssey*, is a popular example; HAL's rules designed to protect humanity from misbehavior lead to a contradiction where HAL concludes he must kill a spaceship's crew.

Since this book emphasizes the accelerating rate of change that software enables, one might ask how far computer intelligence can be taken. In more general terms, is "artificial intelligence" of machines in our future? Are they likely to think like humans, develop egos, and take actions that threaten us?

John McCarthy, the influential technology pioneer, coined the term Artificial Intelligence (AI) in 1956 and defined it as "the science and engineering of making intelligent machines." Since "intelligent" is used in the definition, his definition avoids the challenge of defining what is meant by "intelligence."

At one extreme, the goal of artificial intelligence is to create a machine indistinguishable from a human other than in form. For example, Gerard O'Regan in his book, *A Brief History of Computing*, introduced a chapter on AI as follows: "The ultimate goal of Artificial Intelligence is to create a thinking machine that is intelligent, has consciousness, has the ability to learn, has free will, and is ethical." One could ask whether having a computer exhibit "free will" is in the interest of society, among other questions such as whose ethics it would exercise. In any case, most researchers who say they are working in AI today don't claim to have such extreme objectives.

Nevertheless, the question of whether a computer can exhibit humanlike intelligence has fascinated researchers. Alan Turing famously proposed his "Turing test" in 1950, a test of whether a computer had achieved that goal. In the test, a person communicates sight unseen with two respondents, one a person and the other a computer, and an observer attempts to tell which is the computer and which the human. The test is conducted by text communication only, so speech is not part of the test, perhaps reflecting the times in which Turing wrote his proposal. The human must be honest, and the computer can lie. Thus, if the tester asks, "Do you have two feet?," both would presumably say yes. If the tester cannot tell which is human and which is the machine, the machine passes the Turing test and exhibits intelligence. Turing believed that a computer would someday pass his test, and this view has sometimes been characterized as "strong AI."

The philosopher John Searle argued that the Turing test wouldn't prove anything about whether the computer understands the inquiries. He used a "Chinese Room" argument. He assumed that the method the computer used was equivalent to looking up the question in the Chinese language in a database and pulling out the answer to that question stored in the database in the Chinese language. His Chinese Room argument supposed that written questions and answers in the Chinese language on tiles were fed in and out of a slot into a room where a human who spoke only English was simulating the computer. The human uses the same matching method the computer would to find the answer file matching the question tile, without understanding the Chinese characters on the tiles, and pushes the answer tile through the slot. To the person posing the questions, the Chinese Room seemed to be displaying human intelligence. But the human would have no idea what either the question or answer *meant*. Thus, Searle concluded, passing the Turing test would not indicate any "understanding" or "intelligence" in a computer doing the same thing, simply finding the matching answer to a question. Searle's argument applies to the "intelligence" presented today by computers, since that technology, as we discussed in the section on algorithms, is essentially a statistical model of data, in effect an efficient way of looking up the answers that best match an input pattern.

The idea Searle was conveying is exemplified by applications such as mobile personal assistant applications that allow speaking or typing "natural language," that is, in a way we would communicate with a person. Apple's Siri, for example, is infamous for clever answers to questions we would not normally ask a computer. According to Walter Isaacson's biography, Steve Jobs, at his last Board meeting (where he resigned as CEO), asked Siri whether it was a man or woman. It replied, "They did not assign me a gender." When asked, "What is the meaning of life?", Siri gives varied answers, including a literal dictionary definition: "Life: The condition that distinguishes animals and plants from inorganic matter, including the capacity for growth, reproduction, functional activity, and continual change, preceding death." It also at random provides humorous answers:

> "Life: a principle or force that is considered to underlie the distinctive quality of animate beings. I guess that includes me."
> "To think about questions like this."
> "I don't know. But I think there's an app for that."
> "Try and be nice to people. Avoid eating fat. Read a good book every now and then, get some walking in and try to live together in peace and harmony with people of all creeds and nations."

Hopefully, no one believes the computer is inventing these responses. Much as in Searle's Chinese Room, the question was anticipated by Apple staff and the answers created by clever humans. The software has no idea what the question or responses "mean."

Perhaps another difficulty with the Turing test is that computers can do things quickly that humans can't do easily if at all. Presumably, the human wouldn't have access to a computer or calculator during the test nor the computer to a human for advice, or the test would be meaningless. Suppose the questioner in the Turing test asked for the product of two large numbers, and one respondent quickly gave the right answer. It should be apparent which was the computer. The computer, to emulate the human, must delay its answer to this math problem and perhaps make a mistake. Is this the kind of behavior we would want to have to program in artificial intelligence software just to pass a Turing test?

IBM's "Deep Blue" chess-playing software defeated world chess champion Gary Kasparov in 1997, and this event is often cited as an example of artificial intelligence overtaking human intelligence. But there are only a finite number of moves possible at one time in chess, and, while Deep Blue used a more sophisticated approach, it is possible for software today to look at almost all the moves ahead for at least some number of moves, scoring the position after those moves to decide which immediate move will give the best future result. Software can also store in memory a large number of previous games by grandmasters and match situations that occurred before, pulling up a response that was previously successful with no analysis at all, simply memory. Chess is a game where there is no random or psychological element (unlike poker, for example) and well suited to being over-powered largely by the increasing processing power and memory of computers. This is not an example of computers exhibiting human-like intelligence. One could even consider a human playing chess as trying to emulate a computer, looking ahead as far as possible and studying chess openings to use one's memory rather than analysis to get off to a good start. (This is why the first few moves in chess tournaments tend to occur very quickly.)

Whatever defines human intelligence, understanding human intelligence is a separate goal from creating useful software that emulates some human characteristics. Meeting limited goals such as speech recognition by computers doesn't require we achieve that goal by emulating the methods of humans, as previously discussed. Statistical methods, such as those used in pattern recognition, extrapolate from labeled examples. They don't attempt to model how those examples were labeled. The statistical approach drives technology such as Optical Character Recognition (OCR) and speech recognition today. Statistical pattern recognition has today been applied in an amazing variety of tasks, including difficult tasks such as face and gesture recognition from camera images. This range of applicability (as well as my own history in applying statistical pattern recognition methods to a variety of applications) has convinced me that one could make substantial progress in having a computer doing things that humans even have difficulty doing without jumping the high hurdle of emulating how humans think.

To make this difference more concrete, let's consider an aspect of how today's speech recognition systems convert speech-to-text, a problem I focused on for more than a decade. One aspect of this task is deciding what to decide when encountering words or phrases that sound alike, like "to," "two," and "too." Statistical Language Models (SLMs) are used for this task. An SLM records the fact that "too much" is more likely a pair of words than "two much" or "to much." Similarly, "to go" is more likely than "two go" or "too go." SLMs are used to represent the probability of specific combinations of words, usually triples of words. The statistics are compiled from huge volumes of text sentences. SLMs can make the speech recognition seem very "intelligent," but, obviously, the computer doesn't understand what "too much" means.

In short, the methods used in speech recognition do not attempt to mimic directly the way humans recognize speech or understand language. The methods are developed by finding parameters that optimize performance against example data—empirical methods. Of course, humans learn from examples as well, but the way the brain absorbs and retains this information is certainly different. For example, instead of analyzing millions of sentences at once, we deal with one at a time.

My view is that AI is not a particularly useful way of expressing what we want to do with computers. *We want computers to help humans, not emulate them.* How does a machine understand the subtle implications of a statement like Shakespeare's "That which we call a rose by any other name would smell as sweet" in the context of Romeo's love for Juliet, but her being a member of the wrong family? Or understand that the statement reflects his love for Juliet? We develop some understanding of what it means to be human by growing up in a human body and interacting with other humans in both pleasant and unpleasant ways that wouldn't be practical for a robot to experience. That level of understanding of language may be an interesting research goal, but should it be the goal of commercial products?

On the other hand, if we understand how the neocortex, the part of the brain where what we call human intelligence resides, works in detail eventually, wouldn't a computer using those techniques show human intelligence? Blakeslee and Hawkins in *On Intelligence* countered that argument: "the human mind is created not only by the neocortex but also by the emotional systems of the old brain and by the complexity of the human body. To be human you need all of your biological machinery, not just a cortex. To converse like a human on all matters (to pass the Turing Test) would require an intelligent machine to have most of the experiences and emotions of a real human, and to live a humanlike life."

Understanding how humans think is a worthwhile research objective in itself. Other than the direct value of adding to human knowledge, any breakthroughs in this area are likely to lead to methods that could be executed on a computer for more specific purposes than the full emulation of a human's knowledge and reasoning.

Nevertheless, we are best served when we recognize the difference between computer processing and human processing, and take advantage of what the computer does best. Computers do some things well beyond human capabilities, e.g., storing maps that include almost every street in an area—with little danger of forgetting what they have "learned." Computers can compute the shortest driving route between two points hundreds of miles apart in a few seconds, in some cases taking into account current traffic congestion, and draw a map in seconds to show us the route. Navigation systems can even appeal to satellites to tell us where we are. Computers excel in their ability to analyze large databases and come up with the most consistent explanation of that data, and then to use that explanation to produce information in a form useful to humans. If a computer were limited to human abilities, it couldn't do that.

As these examples suggest, computers do many things humans can't. Let's call the capabilities that can be accomplished by software on a current or foreseeable digital system *Computer Intelligence* (CI).

To be clear, CI often requires human intelligence as input to its analysis. Humans realized that patterns in fingerprints such as "loops, whorls, and arches" could be used to summarize an otherwise complex pattern. Later, computer scientists used mathematical representations of these features in fingerprint classification software. Human understanding that speech is composed of a finite number of phonemes strung together was used in creating today's speech recognition systems.

In some cases, things previously considered the sole domain of humans are actually done better by computers. The spelling and grammar checker in the software program I'm using to write this book is constantly suggesting corrections to my spelling and grammar.

The "science fiction" view of computers is that they will attain intelligence that rivals humans, become smarter than humans, and even become a threat to humanity. Crossing that science-fiction threshold has been called reaching the "singularity" by some futurists, such as Vernor Vinge and Ray Kurzweil. Vinge summarized in a 1993 article, "Within thirty years, we will have the technological means to create superhuman intelligence. Shortly after, the human era will be ended." The problem that I have with this concept and "artificial intelligence" in general is that I don't believe computers have an independent life from humans. They already exceed human capabilities in some areas, such as having a perfect memory. But they count on us to "feed" them with electricity and instructions. We can unplug them at any time; some readers have probably turned over an old PC to a recycling center. Nathan Myhrvold made a telling point in a previously referenced source: "We may be able to make computers as smart as ourselves . . . in which case they're going to make a hell of a lot of stupid mistakes. Just imagine the AI equivalent of a frat house."

A recent picture I was sent of my two- and four-year-old nieces, each intently playing with a tablet computer, reminds me that today's children will grow up with computer technology. They will find it intuitive to connect with computer intelligence as they grow older and will view that connection as something that gives them pleasure and helps them.

Let me provide one more example of this human-computer synergy. In writing this book, I was significantly aided in my research by electronic book readers that let me highlight key observations by authors as I was reading. The software I used allowed me to access those highlights at my desktop computer, despite having marked them while reading the material on a tablet computer reclining in an armchair. Digital review of the highlights was much more efficient than paging through a printed book for pages marked by sticky notes (my previous practice). And, given the large amount of research required for this book and the wide-ranging topics, I found it reassuring that I didn't have to count on my memory to recover the main points the authors were making.

The digital book software isn't replacing my intelligence. To the contrary, it is enhancing my memory and providing me access to what my human intelligence thought was important. Books in general provide a connection to the intelligence of other humans—the authors of the books. Tools that make it more efficient to highlight insights one wants to remember can make this connection more long-term. This example hopefully illustrates the difference between computers telling us what to do and helping us by doing what they do best.

CI is a more useful concept than the ambiguous AI. CI can continually expand its capabilities and accessibility to humans through continuing hardware and software innovation. CI has the potential to enhance the human experience if humans can use CI to add to their intrinsic reasoning capabilities. Navigation systems, search engines, and language translation software, for example, allow us to do more by augmenting our abilities. CI even lets us communicate with our friends and relatives more easily, even when they are in different cities.

In another way CI aids humans, we trust computers and relevant software with many safety functions. In an insightful article by Philip E. Ross in *IEEE Spectrum* (December 2011), he points to this evolution. He notes as an example that the military is examining the creation of helicopter drones to evacuate wounded soldiers quickly, which might allow rescue attempts that entail too much risk for a pilot. The article notes several examples of where computers have replaced humans in tasks involving safety:

- There is no longer an elevator attendant taking us from floor to floor, yet we don't panic when the door closes.
- Laser surgery on eyes is guided by computer, as well as some prostate surgery.
- Hot ribbons of steel are guided through rolling mills by computer.

Commercial airliners are flying by autopilot for most of a flight, despite the presence of a pilot and co-pilot. The CEO of Boeing predicted in a talk in 2011 that we would see planes that fly freight without pilots eventually.

In saying that our connection to computer intelligence is an important and key trend, I must emphasize what I am *not* saying. I don't believe that the capabilities of computers and software will grow exponentially simply because computer technology may increase in power exponentially and computer processors will grow in numbers. I raise this point because a number of authors have used the argument that Moore's Law, which predicts an exponential growth in the number of transistors that can be put on a chip, will result automatically in Computer Intelligence growing at an exponential rate. Some authors argue that the exponential trend in computing power alone will correspond to the computer eventually taking almost any job done by humans today, simply because there will be so much computer power. For example, one sees this basic claim of equivalence discussed at length in Martin Ford's *The Lights in the Tunnel: Automation, Accelerating Technology and the Economy of the Future*.

I certainly agree with the danger of automation taking too many jobs. It is a theme of this book that software will do more, and accelerating automation is a key economic issue that we must address. But the simple argument that exponential growth in computer power will automatically make those devices eventually do anything humans do or that progress in CI will be exponential is simply wrong. This is a bit like saying that increasing the horsepower of every car and the numbers of cars will help us get places quickly and eliminate the need for mass transportation. As every driver can attest, the limit on the speed of getting somewhere is not horsepower, but traffic, signal lights, and highway capacity.

To take a more digital example, the rapid growth in number of smartphones and the increasing processing power in each generation don't cause exponential growth in data rates or clarity of speech transmitted. Similar to the traffic situation, limits such as the data rate supported by the wireless service providers is the limit on how fast one can download a video, irrespective of how fast the processor is on the device or in the server delivering the video. And there are limits on available spectrum for wireless transmission, similar to the limits of highway capacity for vehicles. Further, faster processing might simply use up battery power more quickly or overheat the phone without compensating benefits.

Computer intelligence is limited by the application as well. The spelling checker in my word processing program on my desktop computer notes a spelling error a fraction of a second after I finish typing the offending word. There is little additional utility in it noting the error ten times faster—I wouldn't notice the difference. And I don't see any limitation in the accuracy of the spelling correction with today's technology that would benefit significantly by more processing power or memory. Exponential growth in computing power doesn't add anything when the task performance reaches a certain acceptable level.

To take another example, the navigation system in your vehicle or smartphone has a map that covers the area you are driving. More memory wouldn't add any functionality once that map fits into a navigation system or smartphone. The software computes the best route between two addresses in a few seconds. Computing it one hundred times faster wouldn't change the basic functionality of the navigation system.

As for the size of memory, once a device or service has all the songs that are available for playing or download in its database, increasing the memory size doesn't add value. The limitation in the number of songs available is how fast humans can write and produce songs people want to hear. Increasing memory size by a factor of 1000 isn't going to increase the number of songs by a factor of 1000.

And when it comes to advanced functionality that requires algorithm development, such as natural language understanding, the limitation isn't necessarily computer power—It is invention. One has to tell a computer how to do something before it can do it faster.

Technologies such as "machine learning," "speech recognition," and "natural language understanding," for example, are sometimes misunderstood to mean the computer is "thinking" like a human, and thus they can become smarter if the "brain"—the computer—runs faster. These methods today are statistical in nature; they extrapolate from examples provided by humans. The improvement is more related to the size of the collected databases than the processing power of computers, since this processing is done during *development*, not *use* of the resulting algorithms. Development doesn't have to be done in real time, and it doesn't matter much if the development software runs for days rather than minutes. While improving computer technology certainly helps add to the power of computer intelligence, that intelligence isn't directly proportional to computing power.

This discussion of the difference between what computers can and should do for humans and what they would have a hard time doing is a key point in the later issue of what the acceleration of technology development means to the economy, to what jobs computers can do and those that require humans. The difference in computer versus human intelligence can help us understand where the human-computer connection can make humans more capable of competing with full automation. This is a subject I will dig into more deeply in the later discussion of software's role in economics and jobs.

4

The Nature of the Human-Computer Connection

Given the objective of making computer intelligence more accessible and valuable to humans, how do we do so?

Evolution of the user interface

The first sections of this chapter discussed the evolution of human-computer interaction and the differences between computer intelligence and human intelligence. This section will discuss aspects of user interface design and evolution, introduced briefly in the last chapter.

You deal with human-computer coupling in everyday activities using a personal computer or a mobile phone. Using software on a personal computer is one of the most complex user interactions with technology. Something we do most days, such as checking our email, has so much technology behind it that I will not attempt to list it. Such interaction with PC software is so familiar that it is taken for granted.

Almost any software package you use and the operating system on which it runs has hundreds of features, and significant thought was put into how the user finds and interacts with those features. Some software today has become so feature-rich, however, that there are courses in community colleges in using packages such as Microsoft Office and the Adobe Creative Suite. This aspect of our coupling with software is so familiar that we don't think of the synergy until something doesn't work as expected or we can't find how to perform a specific subtask. We may address the PC in unkind terms in those situations, but fortunately it is not designed to take offense.

There is both art and science in designing user interfaces so that they are more intuitive and so that you have less frustration. There is always tension between the utility of new features and the familiarity of old features—change has an upside and downside. While this isn't the forum to dissect user interfaces in depth, this conflict between change and familiarity is a delicate balance that will impact the evolution of software and deserves elaboration. As discussed earlier, the development of modern PCs, smartphones, and pad computers, as well as game systems and other more specialized devices, has depended on our having learned certain fundamentals of the Graphical User Interface (GUI), e.g., how to use pointing devices and icons.

As devices and applications get increased functionality and features, it sometimes becomes challenging to depend on our intuition. This may be a result of poor design or simply that there are so many features that the designer couldn't make everything obvious. In some cases, the method of finding a feature isn't new, but there are so many features and menu items that finding the feature requires more searching than we'd like. On a smaller device such as a smartphone, the screen size itself makes the GUI less effective, limiting what can be visible on one screen.

Another complication is that companies are using patents on aspects of the user interface to make those interface modes proprietary. There are many on-going patent lawsuits in this area, e.g., Apple suing mobile phone providers using the Android operating system over features of that operating system. To the degree that such suits and countersuits are successful, the consistency of user interfaces between operating systems will be reduced, making learning one after using another more difficult and reducing the consistency we have come to depend upon in user interfaces. The section on patents in a later chapter addresses the issue of software patents in more depth.

In *Trillions: Thriving in the Emerging Information Ecology*, Peter Lucas, Joe Ballay, and Mickey McManus pointed out another aspect of user interface consistency, noting that the WIMP interface created a "community of practice." They explained that such frameworks "encourage a virtuous cycle in which early adopters (usually individuals with particular aptitude or experience) take on the role of first-tier consulting resources for those who come later. As a whole society struggled together to figure out these strange new machines, having everybody trying to sing the same song was of inestimable value." Without consistency, there is danger that these devices will simply be so indecipherable that we end up using only the few functions that we have been able to master. And the technology industry will suffer if we feel we must hold on to older versions of products because we don't want the pain of learning new ones. One example of the impact of inconsistency in user interfaces is the proliferation of remote controls in the living rooms of most homes; most of us use few of the buttons simply because it is a challenge to remember what does what or to set up a "universal remote."

The patent battles, the expansion of functionality in operating systems and applications, and the differences in hardware platforms are challenging the GUI model of a universally understood paradigm. Going back to the user manual or even video tutorials is one option, but one that is minimally effective, since we are usually given information in big, hard-to-remember chunks through these mediums; the expansion of features make this older approach less effective. Help menus provide an option that may provide information in smaller bites, but are more of a symptom of the problem than a solution.

Windows 8, designed to provide a consistent user interface for PCs and tablet computers, to work with either a mouse or touch, is an attempt to provide a solution that addresses some of the difficulties of the classical WIMP interface as the options on the device grow rapidly. The live tiles and other features such as swiping from the sides to get more options are a significant and thoughtful attempt to see how far the GUI can be taken, and to what degree it can be consistent for users on a very wide range of platforms. As I write this, the verdict is still out on user acceptance. Windows 8 is one of those gambles that a relatively large change—even a creative one—won't leave users familiar with the old interface complaining or simply not upgrading.

A hands-free, eyes-free, interactive personal assistant dedicated to helping with an operating system or a particular application could make using new software a much less frustrating experience, particularly if a natural-language interface were used. It would be natural to expect the assistant to do more of what general personal assistants or search functions do, so there might be pressure to expand beyond the limited help assistance.

While speech recognition was important in gaining adoption of the assistant function, the option to type a request and still get the advantage of natural language processing should be available, making the assistant available when speaking is not a good option. The user manual for the device and much of its software would then ideally be "Just say or type what you want." If the user request is unclear or incomplete, an ideal assistant would ask for the information it needs to complete the request. If we remember that we are talking to a computer that doesn't really understand us and help it out a bit in formulating our requests clearly, the technology will become impressive over time.

More generally, many personal assistant apps will have a personality, like Siri with her humorous answers to inappropriate questions. I expect someone will come up with a psychotherapist app that pretends to understand you and gives you encouragement. (It's actually been done; a program called ELIZA was written at MIT by Joseph Weizenbaum in the mid-60s, using text interaction that convinced many that it was a real psychologist, largely by asking questions based on the last input from the "patient" rather than giving advice. Some personal assistant apps are likely to have the specific goal of entertaining us. (If you like detective shows on TV, how about an app that let you interview suspects and determine who did it!)

We'll see many interesting developments using the assistant model, and it has a good chance of helping us adapt to faster change and an explosion of features in software. It moves in the direction of a consistent user interface that is applicable to any digital device.

The role of mobile devices

The International Telecommunication Union put the worldwide penetration of mobile phones at almost six billion in 2011, a remarkable 87% compared to the world's population. There is high penetration even in low-income countries; the penetration in China and India is estimated to be near the world average. Some countries are reported to have more mobile phones than people, including the US, Indonesia, Brazil, and Russia.

Research firm eMarketer estimated in July 2012 that by the end of the year there would be nearly 116 million *smartphone* users in the US, as well as nearly 55 million tablet users. A survey by Yankee Group about the same time found that over 60% of those surveyed under age 34 in the US used their mobile phone as their primary phone. NPD Group estimated in 2012 that almost 40% of children aged 4 to 14 have a mobile phone. The Pew Internet & American Life Project in a survey conducted in July-August 2012 found that 25% of all US adults have a tablet computer. These numbers are even more remarkable since Apple introduced the iPhone only a few years ago (in 2007), launching widespread interest in the category that sparked this growth. The first iPad was launched in 2010, driving interest in this category.

There's obviously something remarkable going on here, something that reflects a cultural shift. It is a cultural shift on top of a cultural shift, driven itself by our increasing dependence on computing and Internet technology for information, communication, and to complete many tasks efficiently. Connected mobile devices give us access to the increasing information resources and services of the Web. We are becoming increasingly dependent on devices we can always have with us for some aspects of memory. Phone numbers, email addresses, and calendar entries are obvious examples, but the list is growing.

And this coupling with computers lets us be more connected with other people through multiple ways of communicating, even when those people aren't immediately available. Mobile devices let us take this connection with computers and people with us wherever we go, becoming more tightly coupled to this ultimate multipurpose tool. These mobile devices make our connection with computer intelligence always available to us.

What we see today is just the start. Increasing computing power in these devices and development of improved algorithms for speech recognition, natural language interpretation, and knowledge representation—some of which will occur in the network—will simply expand what we can do and how easy it is to do. Searching *on our device* as well as the Web with these natural interfaces will become important as we store an increasing amount of personal information, music, books, and photos on the device. And reliable and high-data-rate connectivity almost everywhere will become a consumer and business requirement. Increasingly, we will view our mobile devices as an extension of ourselves.

Another aspect of mobility that isn't directly tied to portable devices is *time mobility*. Technology gives us more flexibility in *when* we do something than in the past, when we had a fixed time window. A phone call pressured us to answer it when the phone rang before voice mail gave us an option of delay without completely losing the call. Email and text messages can be dealt with as time allows. More "live" TV shows are watched delayed on a Digital Video Recorder than are watched when first aired. If we miss a movie in the theatres, we can watch it after it leaves the theatres on many devices. Mobile devices are part of this trend; in addition to giving us flexibility in *when*, they add flexibility in *where*.

Even classical web searches are evolving to better serve our needs; the leader in web search, Google, has launched a Knowledge Graph as a way of displaying something closer to the answer to a search, as opposed to just a list of web sites. The Knowledge Graph uses semantic processing—knowledge of what words *mean* in addition to using them as matching keywords—to group together related knowledge about a topic. If the immediate results don't answer the inquiry, then one can of course drill down. At this writing, Google doesn't differentiate between search and "personal assistants." One can speak a request into a Google search box as an alternative to typing it; a company blog in July 2012 said that Google considered natural-language voice search simply part of search and wasn't planning to call it an "assistant" or give it a personality. One reason for this attitude could be that people have learned how to phrase a typed search request telegraphically to get the results they want, leaving out extraneous words. Google's search engine is currently optimized for that case, and it's likely Google wants time to adapt to inquiries posed in more verbose natural language, although they are offering that option.

In another sign of the evolution toward getting to the result you want more quickly when interacting with a mobile phone, company customer service apps on mobile phones are beginning to proliferate, albeit slowly at this writing. Some offer natural language interaction through voice or text and display the requested information (bank balances, recent checks cleared, etc.) on the screen in response to the request. I've argued in a blog and my newsletter that every company will eventually be expected to have a customer service personal assistant just as they are expected to have a web site today.

If Web search hasn't changed your habits about where you begin if you need information, then I'm surprised. This is but one in a chain of changes of habit that software has brought. You might be annoyed if a business associate, friend, or relative didn't have an email address. And don't you take more photos now that they are digital and a bad photo doesn't waste film (if you are old enough to have ever used film)? And doesn't having a camera always with you in the form of a mobile device make for some wonderful extemporaneous images?

Now imagine this type of change accelerating. With hundreds of thousands of apps for smartphones, clever inventors can quickly deliver to you new things you can do with your mobile device. The delivery of these apps has been accelerated by another software innovation, the "app store." By simply publishing an app in an App Store of one of the providers of smartphone environments, a developer has an automatic mechanism for a consumer to find the app, purchase it, and have it delivered.

And the devices will continue to evolve. If you bought a new PC every five years to keep up to date, you may find yourself buying a new smartphone every two years when you can get a discount from the service provider. An early chapter listed trends that are accelerating what software can do for individuals and businesses. And the increasing computer power and continuing innovation allows tighter connections with you and more reasons for you to connect.

And the form factor of mobile devices will continue to evolve. I've mentioned Google's wearable eyeglass display and Motorola's headset computer with an attached display, both with hands-free control by voice and motion. Many people use Bluetooth earpieces to talk on mobile phones, and thus are able to talk to voice personal assistants without touching the phone if the software is properly set up. Automobiles are evolving toward the "connected car" with voice control of entertainment sources and connectivity to mobile phones. The advantage of the mobile device is that it is always there—one doesn't even have to take a device out of one's pocket or purse when using a wireless microphone. The technology becomes part of us in a more obvious way than, for example, a keyboard that remains on our desk.

Personal assistants

A key aspect of this evolution of mobile devices is the Personal Assistant (PA) that I have mentioned many times. To be more precise in the use of that description, a PA is software that can take a communication posed in a natural language as speech or text, interpret the desired intent of that communication, and provide the user with the desired result as directly and accurately as possible, possibly clarifying the request by dialog. The result can be presented by voice or other means available on the device, including text. To the degree that there is context or other input or output modalities (e.g., touch or graphics) available that can help get or deliver the desired result, the personal assistant may use those resources. The personal assistant may also proactively volunteer information that is relevant to the current situation of the user.

The last point is currently a feature of some standalone capabilities today. For example, Google Now! will do such things as inform you of flight status when you have an airline reservation or of an upcoming appointment. Microsoft's "live tiles" in Windows 8 change to show information such as the current weather forecast or new tweets.

While it isn't a current feature as this is written, it would be a natural evolution if one personal assistant application could be supported by another personal assistant application through an appropriate transfer of control. A specialized PA may only deal with a restricted context, such as customer service for a specific company. Ideally If requested from a general PA or another specialized PA, e.g., by saying "XYZ Company customer service," the active PA can launch the company customer service PA. Ideally, the customer service PA will return the favor by a means of returning to the launching PA or a managing (general) PA. PAs can be specialized or general. A general PA, such as Apple's Siri, attempts to deal with almost anything the user requests, defaulting to a reply such as "I can't help you with that" when it can't handle a request.

This movement between PAs can be much like transfers between web sites through clicking on a link. It would be ideal if a common standard emerged to allow such transfers to be done easily. Like web sites, this might require a registration process so that PAs had unique names, like unique web addresses. If so, some semantic equivalents might also have to be registered so that a user didn't have to guess the exact wording. For example, "XYZ Incorporated assistant" might be equivalent to "XYZ personal assistant" and variations. (New companies are advised to pick a multi-syllable name to easily distinguish the company for speech recognition systems.)

The reason for describing in detail what I mean by a personal assistant is that a Personal Assistant Model (PAM)—a useful shorthand for the functionality I've described—is a significant innovation in user interface design. The PAM has the potential to have the same revolutionary impact as the early Graphical User Interface had in the adoption of personal computers by businesses and the general public. From the point of view of the focus of this book, it allows the tightest connection between people and computers of all user interface innovations, as well as full integration with existing user interface components such as the GUI. The language processing technologies (e.g., natural language interpretation) have a way to go to make using a PA entirely a natural experience, but I believe the technologies have passed the tipping point of utility that will encourage users to adopt them, and that the technology will continue to improve.

It's not just a future trend. A survey of 1,000 U.S. consumers published by Nuance Communications in January 2013 found that 75% stated that they either have their mobile device "always on them" or "at hand." And 90% of those who have a personal assistant feature on their phone say they use it because it's quicker, easier, and more convenient than traditional mobile use, with nearly 60% using it every day. The most common uses of mobile assistants reported were for driving directions, the weather, and restaurant recommendations.

In a 2012 entry in my blog, I introduced the idea of *Ubiquitous Personal Assistants (UPAs)* to emphasize that the concept of the PAM would lead to a desire for the *same* personal assistant to be available through multiple devices, retaining data and personalization learned on one device for use on another. If an assistant application seems like the same assistant on multiple platforms, not just on one mobile device, it makes the user interface to a partial extent independent of the device. Ubiquitous availability can make a personal assistant application a safe extension of our human abilities that is available wherever we are and whatever digital device we are using.

Making full and efficient use of a PA requires providing the assistant with personal information—our contacts, our schedule, where we are, past preferences and interests, and more. We've grown accustomed to at least some of our computing devices—PCs and laptops in particular—containing extensive personal information, e.g., past emails. We depend on this history and reference it frequently. On mobile phones, we have at a minimum contact phone numbers and text messages.

The growing importance of these resources to individuals will drive desire for integrated, easy access to personal information and preferences stored on multiple devices owned by an individual and information on personalized web sites such as Facebook. A UPA could be a consistent connection to these resources. A UPA would most likely also be a general personal assistant, adding access to more general information sources such as business web sites and map services and accessing specialized personal assistants when needed.

If a UPA is the main user interface we associate with all the digital devices connected to the Internet, the company providing the UPA, whatever the device on which it is used, has tremendous power and potential for revenues (from advertising, in particular). Some companies (Apple being the most likely) will probably try to leverage this connection with the user to encourage them to buy the hardware from that company. Others (like Google, Amazon, and Facebook) are likely to try to support platforms from other vendors. Microsoft has signaled a cross-device strategy built on Windows. And there are "independents" such as Nuance Communications that will license core language technology to companies and could be in essence a provider of private-label personal assistants; they are already offering a speech-enabled personal assistant "Nina" that can be adapted to a company's products (and renamed). There will be other players, but I do believe that there will be a limited number of "communities" that most companies serving consumers must join, as discussed in a later section. This trend obviously has significant implications for the economy and for individual companies. The theme of software accelerating technological progress suggests that this trend may occur faster than one might expect. Companies that don't examine their role in communities may find themselves playing from behind on the scoreboard.

Apple's iCloud recognizes the desire for ubiquitous access: Apple summarized (on a web site in October 2012): "iCloud does more than store your content—it lets you access your music, photos, calendars, contacts, documents, and more, from whatever device you're on. And it's built into every new iOS device and every new Mac." Of course, there are independent services such as Dropbox that allow storing files in the cloud and accessing them from many devices. But Apple's apparent vision is making that information more automatically available—as long as you buy Apple products.

A Ubiquitous Personal Assistant might be viewed as expanding the concept of search. We have separate search on our PCs for information stored there, search on the Web for specific information, search on individual web sites for information within the site, search for contacts on our mobile phones, search within specific applications such as email, and many more pockets of search. UPAs would ultimately allow search in any of these pockets from one application available on all devices. The application might be considered a single "personal assistant or an extension of "web search" (with "voice search" using natural language as an option). When this book refers to "personal assistants," I mean the functionality described, whatever it is called by the provider.

The complexity of all these sources would encourage a system that "understood" where the information might be, rather than requiring navigation to specific sources. For example, consider a request such as "Where is next month's planning conference?" An application trying to answer this question might interrogate your calendar, your email, and your text messages, consolidating information from multiple sources. A reply might be, "In an email dated June 2, 2012 [link provided], John Doe scheduled a planning meeting for June 21, 3 PM Eastern, at the company's New York offices, 888 Zeroth Street, suite 500."

This trend will inevitably raise privacy and security issues. The convenience of having information available everywhere implies the ability to communicate that information through networks and make it accessible from multiple devices, and thus makes that information more vulnerable to attack. The UPA trend could be slowed or derailed by major security failures, and full expansion of the trend will require companies that offer UPAs to make a major commitment to protecting personal information. However, if the various devices use biometric identifiers such as voice characteristics before allowing access, the risk may be minimal. The privacy and security issues exist with or without personal assistants, as this book discusses in a later chapter.

Mobility has made access to computer power something we can constantly have with us. Ubiquitous Personal Assistants may be the ultimate way we channel that power.

Limitations on the mobile connection

There are hurdles to having a mobile "assistant" always with us. One is the limitation of available wireless spectrum to support the growth in data being handled by wireless devices.

Available spectrum for the growth of cellular networks is limited, and could impede growth of some services or raise their price to the point that it deters usage. The most evident trend addressing this problem is the spread of WiFi networks connected to the Internet, local wireless networks that have limited coverage, but are becoming more common in homes, coffee shops, airports, and similar venues. This trend may need to be encouraged by government policies or innovative services that aggregate such resources so that they become automatically available to mobile devices at some affordable subscription fee without a lengthy log-on or payment process at the site.

A similar problem as our mobile devices become more powerful is battery capacity. If these devices become almost part of us, it becomes almost a physical disability if the device runs out of power.

In the long term, innovations in battery technology could reduce the problem. In the shorter term, recharging connections in automobiles are likely to become a standard feature, with a wired USB connector or docking station also providing a connection to the electronic systems in automobiles that today often offer only Bluetooth wireless connections to mobile devices (or none). Another good option is wireless charging (induction chargers that work like transformers using varying magnetic fields to transfer energy without a wire). If the receiving end were built into the mobile device, it could be simply placed on a charging pad at your desk, a coffee shop, restaurant, or on public transportation. Some mobile phones already have the ability to use a wireless charger.

Implications for business of the tightening human-computer connection

The importance of software being coupled to people more tightly was signaled in a remarkable annual letter to shareholders by Microsoft CEO Steve Ballmer, posted on the Microsoft web site in October 2012. Ballmer emphasized that a strategic change in company policy focused on tighter person-to-computer interaction and was required to match trends both in the consumer and business space. Ballmer noted, "Over time, the full value of our software will be seen and felt in how people use devices and services at work and in their personal lives. This is a significant shift, both in what we do and how we see ourselves—as a devices and services company." He noted that enterprises were increasingly driven by consumer trends—"the Consumerization of IT," he called it—and Microsoft was going to adjust to that trend.

Ballmer listed five "focus areas." In his words:

- "Developing new form factors that have increasingly natural ways to use them including touch, gestures and speech.
- Making technology more intuitive and able to act on our behalf instead of at our command with machine learning.
- Building and running cloud services in ways that unleash incredible new experiences and opportunities for businesses and individuals.

- Firmly establishing one platform, Windows, across the PC, tablet, phone, server and cloud to drive a thriving ecosystem of developers, unify the cross-device user experience, and increase agility when bringing new advancements to market.
- Delivering new scenarios with life-changing improvements in how people learn, work, play and interact with one another."

Ballmer's points heavily focus on usability and a tight connection between computers and people. Software increasingly must connect with people in a natural intuitive way, including speech recognition and machine learning that not only learns from what people do with software, but anticipates what they might want to do based on what people in similar conditions did. That he considered this trend a major strategic shift for Microsoft emphasizes the difference between today's game-changing trends in software versus its historical evolution, in the past paced by periodic delivery of new "versions" of software.

In a later interview published in the Wall Street Journal on October 30, 2012, Ballmer was asked about his mission for the company. His response was: "We talk internally about enabling people and businesses to realize their full potential. Whether that's sexy or not, putting you in control of your world—and letting you do things you weren't able to before—that's what we're about."

I mention Ballmer's courageous attempt to signal deep change in one of the world's largest and most influential companies and his apparent focus on the connection between people and computers because it is one of the clearest signals that something is going on in software development that can change the economy significantly. Microsoft's strategy in part reflects actions by competitors. The competitors are all attempting to create ecosystems that support a family of devices, software, and supporting services in the network (the "cloud"). To a large degree, they are looking to tie customers and companies to their specific vision of the evolution of computing.

The "consumerization" of Information Technology that Ballmer referred to is driven by individuals being so tightly dependent on devices such as mobile phones or tablets that companies can't forbid their use in company activities. This trend has been labeled Bring Your Own Device (BYOD), and it is creating headaches in IT departments as they try to maintain the security of trade secrets and business strategies in such an environment. More fundamental than the headaches, however, is the improvement in communication and productivity that these devices allow.

Beyond companies internally taking advantage of and adapting to the tighter coupling of people and software, all businesses must adapt to the new reality in their connection with *customers*. Consumers are fast-forwarding through TV ads and largely ignoring display ads on web sites. Display or intrusive ads on mobile phones are particularly annoying in that they take over the small device and interfere with whatever task the consumer is performing.

Personalization of advertising is typically more effective if done in a way it seems to be helping the consumer find something they want. Amazon's suggestions of books and products are tailored to what you bought before, for example. With mobile phones, the suggestion of a business in the neighborhood having a sale might be effective.

The most likely evolution of marketing/advertising is the use of preferred search. If someone searches for something, they are obviously interested in that product or service or something connected to the search. Search engines have put advertised links at the top of search results for a long time, so we are used to distinguishing ads from unsponsored links. With direct-to-content results trying to give the answer more directly, paying to be the first option is likely to be an effective advertising model for the search engines and the advertiser. This must be handled carefully to let the consumer know the result is biased, but this advertising model could become common.

The company personal assistant may be an effective way to connect with customers. Consumers may download a company personal assistant on their own, perhaps motivated by a customer service inquiry. Surveys conclude consumers prefer self-service to talking to an agent when the self-service method, e.g. a web site, is effective. Alternatively, a search function or general personal assistant could call up the company assistant when requested by brand or by suggesting personal assistants that fit the request. (The company providing the general search assistant could possibly receive advertising revenues for this service.) The company personal assistant must be carefully designed and well supported. A human agent may be behind the scenes, typing in responses that are then spoken by the personal assistant in a synthetic voice, or the customer service representative may join the conversation "in person."

The company personal assistant can be taken even further, providing interactive entertainment. The "advertisement" could be woven into the entertainment.

Further, within a company, software that aids in doing one's job may evolve to have a specialized personal-assistant interface option. Some software that allows using voice to update sales databases is already available, for example.

Too much data

Previous sections suggested that we will have increasingly friendly and flexible ways to request information. But how do we manage the huge amount of information that is becoming available to us? If mobile devices are to effectively increase the useful information available whenever we need it, the result of a request needs to be as crisp as possible.

A previously mentioned field that was evolving as this book was written can be called "knowledge representation." It refers to computers helping us with the task of converting raw information into insights and condensing large volumes of text information into answers. It is perhaps the least developed of the advanced information processing technologies needed to meet the full vision of this book.

The intent of knowledge representation is to take very large sets of related data sources and analyze them into a form where a query can provide an answer with little if any further probing. To take a simple example, it should be possible to develop an algorithm that would ignore replications of the same information at different web sites and present the source site (perhaps with an option to see commentary from other web sites).

A search engine that provides the answer desired when possible, as opposed to a list of web sites, is an example of both a trend that is occurring and a challenge to do effectively. I've called this approach

Direct-To-Content (DTC). Today's personal assistants and search engines increasingly attempt to deliver the information requested directly rather than simply a list of candidate web sites that may contain the information. For example, Google's Knowledge Graph will in some cases represent information based on a semantic understanding of the request. Software processing a request for "Marie Curie" would recognize that she is an historical figure and provide summary information about her alongside a listing of web sites with additional information. The search entry "taj mahal" might refer to the monument in India, a Grammy Award-winning musician, a casino in Atlantic City, NJ, or perhaps a local Indian restaurant; the Knowledge Graph will highlight the most common target, but also provide options to search the alternative meanings. Knowing the alternative interpretations of a request could also allow the search software or PA software to ask which alternative the user wanted.

Thus, knowledge representation is the part of DTC that lets an inquiry find a more direct and complete answer rather than just identifying many possible sources. For example, a request for "How do I make chili?" might provide several highly rated recipes for beef and turkey chili, with links to more options (or perhaps a request for clarification, "Beef or turkey chili?"). The obvious commercial value of the answers to requests that involve buying a product or service are creating significant competition between companies such as Google, Apple, Amazon, Microsoft, and IBM to find a popular DTC approach.

I mention IBM in particular because an example of knowledge representation technology is IBM's Watson, the software program that beat two champions on the TV show *Jeopardy*. The key capability of Watson technology is analyzing multiple text databases so that the information they contain can be organized, integrated, annotated by context, and accessed more directly and quickly than just searching text for keywords.

As an example, IBM has created the Strategic IP Insight Platform, or SIIP, for a healthcare application. SIIP scans pharmaceutical patents and biomedical journals to discover and analyze information pertaining to drug discovery. According to the company, IBM cataloged 2.5 million chemical compounds from 4.7

THE NATURE OF THE HUMAN-COMPUTER CONNECTION

million patents and 11 million journal articles between 1976 and 2000. The technology goes well beyond classical keyword search by using semantics, context, and natural language interpretation to create focused results to an inquiry.

Nuance Communications is working with IBM to provide a similar service for enterprises. Nuance's Prodigy is an implementation of Watson-style technology that is intended to take company information sources such as Web sites and internal documents and distill them into an easily accessed information source. One application might be to make "answers" available to customer service agents or personal assistants addressing a customer request.

Knowledge representation is particularly effective when combined with natural language processing technology that allows a request to be stated in a natural language, as if interacting with a person (as either voice or text). Vlad Sejnoha, the Chief Technology Officer of Nuance Communications, gave examples in a company presentation in December 2011 of how knowledge representation could be combined with speech or text interpretation to provide answers to questions. One example was medical: "Is a hormone deficiency associated with Kallman's syndrome?" was answered, "Yes. A deficiency of GnRH is associated with Kallman's syndrome" (with source evidence listed). Another example was the question, "Any movies with George Clooney playing tonight?" spoken into a mobile phone, with the answer "I found three" followed by listing of the movies and nearby theatres showing them. Note that this last example uses the understanding that the requestor would want movies currently showing nearby (implied in part by "tonight") and subsidiary knowledge of where the person is (e.g., because the request was made on a GPS-equipped mobile phone). Only a few years before Sejnoha's presentation, both of these examples would sound like science fiction. In the same presentation, Nuance CEO Paul Ricci emphasized this broad view of natural language processing—both text and speech—as "moving from recognition to outcomes."

Apple's mobile voice assistant Siri uses the DTC approach, trying to provide the response to a spoken request as directly and succinctly as possible. The DTC model is a fundamental change in search. Emphasizing this point, Eric Schmidt, executive chairman of search giant Google, in a letter to the Senate Subcommittee on Antitrust,

Competition Policy and Consumer Rights in 2011, said Apple's Siri was a threat to his company's search business, especially on mobile devices. He wrote: "Apple's Siri is a significant development—a voice-activated means of accessing answers through iPhones that demonstrates the innovations in search . . . History shows that popular technology is often supplanted by entirely new models." He referred to Siri as a "search and task-completion service." Schmidt's comment on the threat to Google's search business suggests of course that the DTC model also has ramifications in advertising, as I suggested earlier.

At this writing, it's not broadly recognized that Knowledge Representation is a key technology and different from "natural language interpretation" as it is normally used. The technology is challenging, but the accelerating pace of software development—as well as the clear financial motivation for success—is likely to make it better and more generally available rapidly.

Increasing information accessibility and evaluating quality

One problem is that the Web and its evolution may be creating an overdose of information. Often, insights or news at one web site are referenced, replicated, or restated at multiple other web sites, sometimes making it hard to find the original source. But replication (if it is accurate) isn't as much of a problem as (1) the sheer amount of information that might be of interest to an individual; and (2) misinformation, wrong information that is either deliberately misleading or simply based on a misunderstanding of facts.

How can we assess the quality of information we find? Basic web search engines work by taking words or phrases and matching them to web sites that seem to provide that content, ranking the web sites by sophisticated algorithms. One method that Google and other search engines use, for example, is to see which sites are most referenced by other sites. Google founders originally applied the idea to technical references—which papers were most cited by other papers—in research while at Stanford University. Search engine methods today go well beyond this basic idea, including some methods for defeating attempts to "game" the system, e.g., by creating fake web sites that reference their primary site.

The amount of information on the Web and within enterprises is growing rapidly, and, since most information is stored digitally today, is accessible by computer. The large amount of information on a given topic can be a problem in itself, presenting a user with so many options to analyze that finding a list of sources is just the beginning of a long task of plowing through potentially contradictory sources. Algorithms that implicitly measure the opinion of others about a source by the frequency of its being referenced or looked at run the risk of its simply having been the earliest source of specific information, being listed arbitrarily on a popular site, or other aberrant effects.

One historical way that information has been selected for reliability is through human filtering. In this model, an expert (or at least a person whose good judgment has been demonstrated by past results or professional experience) provides an interpretation and sometimes a summary of information on a specific topic. An example is a periodically published, paid-subscription, no-ads industry-specific newsletter. In the ideal case, the newsletter writer puts any specific news in perspective, makes the key points clear, goes beyond jargon, extracts from purveyors of the news missing details, and interprets the gist and importance of the information in a larger industry context. The process of one person going through industry-specific developments month after month should lead to insights. Such periodic newsletters, however, are being replaced by the Web's instant news and analysis.

Today, there are web sites and blogs with widely varying standards. Many web sites are constantly updated and often supported by advertising. One could argue that the content on ad-supported sites might be biased by who advertises there. Such sites are often promoted by emailed summaries of news on the site, and there is pressure to post breaking news as quickly as possible, making deep analysis and the journalistic practice of confirming and clarifying the basic news with the source more difficult. In the extreme case, some sites ("mash-ups") just have links to press releases or other sites and don't add much more than gathering what the purveyor thinks is relevant news. This is filtering, but usually without commentary on the source.

Can expert filtering and commentary be retained in an environment where news propagates so quickly and is largely ad-supported? Subscription-based models still exist, with delivery generally digital, and monthly magazines still survive. The trend away from the paid model may eventually reverse as the "noise" level of information increases, although it may be necessary to provide filtered information closer to the date it occurs. Since any one individual cannot practically be available to analyze breaking news every day, the trusted individual is giving way to the trusted organization. While some might argue that "trusted" is a strong term to apply in this case, dailies such as the *Wall Street Journal* manage to charge for both their paper and web versions. Monthly magazines such as the *Economist* have kept readers by focusing on analysis and depth of reporting rather than timeliness (and by also offering digital and even audio versions).

Beyond expert filtering, there are crowd evaluations, where readers can "like" an article or service. This approach is likely to at least flag the poorest content.

There is an unfortunate type of filtering that doesn't select quality content. Some governments control access to selected information (censorship). Google's November 2012 Transparency Report reported that the number of government requests to remove content from Google services was largely flat from 2009 to 2011, but spiked in 2012. In the first half of 2012, there were 1,791 requests from government officials around the world to remove 17,746 pieces of content, Google reported.

Connecting people to people

Hopefully, talking to our personal assistants will not take the place of talking to other people. While a focus of this book is the increased communication that technology is allowing with computers, accelerating the trend of computers becoming aides to our doing what we want to do, I should hasten to reiterate that software trends also are making our connection with other humans easier and expanding our social community.

The key trends in this area are familiar:

- *Direct voice communications:* Landline and wireless voice calls are by now considered old technologies. However, software advances are making voice calls more flexible. Enhancements include services converting voicemail to text so the subject of a call can be quickly reviewed before returning a call, sometimes called "visual voicemail." Speech recognition can allow quickly dialing by speaking a name on a mobile phone—"Call John Doe." Conference calling services allow voice-to-voice meetings of groups, with software adding options such as videoconferences and/ or web-based visual presentations.

- *Indirect communications:* Email and texting were the earliest innovations in allowing us to send messages where we didn't expect immediate responses, and where the recipient could respond at their convenience. Email also allows sending messages to many people at once, without creating the message over for each; of course, this has the downside of spam being easy to send. Tweets using the Twitter service are more recent.

- *Social media:* Services such as Facebook and Google+ allow us to post messages and pictures for our friends and acquaintances to view at their leisure, as well as comment on the postings. Business-focused services such as LinkedIn allow similar relationships in our professional endeavors.

- *Data entry:* Most people wouldn't consider entering data into a computer database a means of communicating with other people, but it is in many cases. Data entered is often used by another person looking up something in the database. Electronic Healthcare Records are a way a doctor makes his observations available to other healthcare professionals, for example. This point of view also suggests that entering an appointment in your calendar is like talking to yourself through a later automatic reminder.

We can expect continuing improvement and features aiding us in using these person-to-person services. Personal assistants already aid in doing so; for example, one can say, "Text to John, Can you meet for coffee at 5, the usual place?," then review the contact info and transcription to text and send it.

Some enhanced communication technology allow us to more easily participate in a group we want to be part of. There is something intrinsic about the human desire to be part of a group, and technology helps.

Sherry Turkle in *Alone Together: Why We Expect More from Technology and Less from Each Other*, views technology trends as isolating us, arguing that there is no substitute for face-to-face communication. She argues, "We fear the risks and disappointments of relationships with our fellow humans. We expect more from technology and less from each other . . . Technology is seductive when what it offers meets our human vulnerabilities. And as it turns out, we are very vulnerable indeed. We are lonely but fearful of intimacy . . . Our networked life allows us to hide from each other, even as we are tethered to each other. We'd rather text than talk."

Turkle is probably accurate in describing one source of the popularity of software-aided communication. But the other side of that coin is that such lower-stress communication lets us keep up with more friends more easily and frequently, perhaps building friendships by *supplementing* face-to-face meetings. Perhaps the lesser tension in an email, text message, tweet, or Facebook posting relative to an interactive conversation allows maintaining a tighter connection with others because of the very factors she cites. Video calls and videoconferences provide an option that is close to face-to-face communications without the travel.

Software can also *enhance* face-to-face communications. The reader may have been in a situation where a smartphone was pulled out to show a photo of a person or place being discussed. Business meetings have long included presentations on laptops or other portable devices to enhance communications. It is hard to convincingly paint technology as a barrier to human-to-human connections.

Robots

Turkle expressed particular concern about our substituting "robots" for human interaction. She cites robot pets as one example of a substitute for human connection (without pointing out that live pets could be a substitute for human connection as easily). Robots are one type of hardware that requires sophisticated software as a key component, and, science fiction has led us to eventually expect robots that are almost human. That vision of robots is perhaps the ultimate image of hardware replacing humans.

"Robots" is one of those terms like "Artificial Intelligence." It conjures up images, mostly from books, movies, or TV shows. One could informally define a robot as a machine that shows some characteristics of a human assistant and that aids people. (The origin of the term is from a play by Czech playwright Karel Čapek, and the Czech word means "forced labor.")

A clear definition of a robot is hard to find; dictionary definitions I've looked at don't seem to fully capture what we imagine when we hear the term. For example, we usually visualize a "robot" to be capable of changing its location; if not, a washing machine might be considered a robot. Whatever the definition, much research is proceeding under the heading of robotics, e.g., creating a mechanical hand capable of picking up an object without crushing it. The speech recognition and natural language processing that we are familiar with on mobile phones, for example, might be considered a way to allow a robot to understand commands and even carry on a limited conversation.

We imagine a robot to be at least partially autonomous, that is, work without requiring a human to control it directly (other than to give it instructions). An automobile is mobile, and has features such as an automatic transmission that previously required specific human actions, but we don't consider it a robot because humans directly control most of its activity. An autopilot flies an aircraft, but isn't considered a robot. What would be the advantage of having the autopilot be a device seated at the controls and using camera "eyes" to read instruments and mechanical "hands" to manipulate the controls?

The excellent 2012 movie *Robot and Frank* takes place in the "near future," according to the leader in the movie, and features a caretaker robot that is more realistic than most robots in movies. However, "near future" may be optimistic. The robot does things such as housekeeping and cooking that, on reflection, are extremely complex operations involving a lot of understanding and manipulation, automation that requires a lot of invention that isn't obvious today. Such operations may be done by specialized devices;

e.g., iRobot sells what the company calls a "vacuum cleaning robot," shaped like a large hockey puck. A single robot doing all the tasks involved in house cleaning is probably impractical, since different tools and form factors would be required for each task. For example, a robot that cooked would probably have a built-in food processor and would probably have to be "handed" the ingredients used in its recipes, rather than expecting it to take them out of the refrigerator and cupboards.

The above is not an argument that robot-like devices won't be serving us in the future. If you take away the requirement of motion, machines such as washing machines and dishwashers are forms of a robot, assisting us with mundane but frequent tasks, but specialized to their function. Ha-Joon Chang, in his book *23 Things They Don't Tell You About Capitalism*, even claims that the washing machine changed society more than the Internet by freeing more women to join the workforce. Even if one doesn't fully accept Chang's conclusion, he points out the strong impact of familiar and old technologies.

As another example of the power of specialization, a device by the company Snuza is specialized to prevent Sudden Infant Death Syndrome (SIDS), the leading cause of death for babies in their first year of life. The Snuza Halo monitors the baby's movement by clipping to the baby's diaper near the stomach. A sensitive motion detector monitors and constantly recognizes the baby's movement, including breathing. If the Snuza Halo does not sense movement within a 15-second time period, it vibrates, a technique used by hospital neonatal care units called "Cutaneous Stimulation," which usually causes movement of the baby. If movement is not sensed within a further 5 seconds of the vibration, an audible alarm is triggered to alert parents. Such highly specialized devices can be more effective because of their form factor and where they are placed than a general-purpose robot "watching" the baby and "touching" it if it doesn't move in a period of time. Dedicating a generalized robot to this task wouldn't be cost-effective.

Given the impact of such specialized machines and the power of specialization, it is likely that general-purpose robots will be more of a research challenge than a commercial reality, no matter how long we wait. It's a bit like artificial intelligence viewed as an attempt to mimic a human. If we want a human, why not give one a job? It looks like it may be a long time before the planet runs out of people to fill jobs available. The problem is more likely to be a shortage of jobs, and this book argues that humans will increasingly have easy access to computer intelligence to aid them.

Software-based devices can certainly help humans perform tasks, but calling such devices "robots" brings up unnecessary connotations. A specialized robot, if that is how it is described, is simply another example of a machine that software evolution can make possible. It is one example of hardware evolution accelerating through rapid changes in software.

5

Security and Privacy

The increasing connection of people and computers raises security and privacy issues. Using cloud-based services often involves company and government storage of private information. Information such as social security numbers and credit card numbers can expose us to fraud if stolen or misused. Our personal information, such as parts of social sites that we have designated for limited access, can be hacked or used without our explicit permission by companies owning the sites to target ads. There have been enough reports in the press for us to understand that both theft of our information and misuse of our private information is a real risk with today's software. For example, there are frequent revelations of the exposure of supposedly safe data at major companies, including credit card data. It's not just personal information at risk; the theft of sensitive technology and business data from major companies is sometimes even attributed to intelligence operations of major countries.

Security issues occur at several levels:

- *Individual:* We are at risk that our computers and other electronic devices will get "infected" with viruses or other harmful software.
- *Companies:* Company databases can be compromised to reveal information on customers or R&D and business information that is considered proprietary. Theft of customer data can harm reputations and trust.
- *Governments/Countries:* "Cyberwarfare" is a serious concern. At one level, it is theft of information that can benefit another country or company. Most seriously, it could be a concentrated attack on infrastructure such as

communication, transportation, or electrical systems, with both economic and safety impacts. A later chapter discusses cyberwarfare.

These are not entirely separate issues. Many successful attacks on business are the result of individual employee's actions using company computers or attaching their own laptops or tablet computers to company networks; hackers who can manage to get their software on an individual's PC or other device can often use it to get software onto company servers or to get to data supposedly behind a firewall. This problem is further complicated by more devices that were previously special-purpose becoming in effect general-purpose devices that can run complex downloaded software (e.g., a smartphone versus a feature phone). Who evaluates in depth the hundreds of thousands of mobile apps available in App Stores today?

Many security failures are often out of the control of individuals. For example, it was revealed in October 2012 that citizens who filed tax returns in South Carolina after 1998 might find critical personal data compromised. An unidentified foreign hacker was reported as having penetrated the state's Department of Revenue, stealing about 3.6 million social security numbers and 387,000 credit and debit card numbers. South Carolina Governor Nikki Haley described the severity of the situation: "The number of records breached requires an unprecedented, large-scale response by the Department of Revenue, the State of South Carolina, and all our citizens." While this is an extreme case, it certainly isn't an isolated case.

Security issues are one aspect of the "soft" in software. While ease-of-change can have advantages in improving software performance over time, it has the downside of also allowing less constructive changes without the nefarious intent being obvious.

The issue is not a simple one. In the South Carolina case, if your data was stolen, you don't even have the option of no longer doing business with the entity, as you could with most companies. The later chapter on cyberwarfare discusses some potential long-term approaches to reducing the problem.

Privacy

Most loss of privacy isn't due to misbehavior; today, much is voluntary through social networks, blogs, etc. One of the key issues going forward is a tension between allowing software to help us and preserving our core privacy. The key factor here may be "connectivity," that is, the connection of many devices to remote servers somewhere in the "cloud" through the Internet or a wireless connection. Many services require connection to information sources that cannot be on a local device because they change too often. Some services, such as advanced speech recognition and natural language interpretation may require more computation than is available on the device. Personal assistants allowing voice interaction do most of the processing and interpretation of the speech in the network.

The processing in the network also allows speech recognition and natural language processing to be updated constantly. For example, new addresses and businesses are created constantly, and recognizing a new business name requires updating the speech recognition and natural language processing to understand a request involving that business name. Even if the devices eventually have the capacity to hold all this information (and the battery life to support such an extreme case), it's not practical to download updates of large databases to hundreds of millions of devices daily. Thus, a network connection, with the risk to the privacy of data flowing through it, is a practical necessity even as connected devices get more powerful.

Further, the desire for Ubiquitous Personal Assistants, assistant applications that seem the same when invoked on different devices, requires a cloud-based service that retains and synchronizes information. Since this information should be available only to one's personal assistant software, it shouldn't create privacy issues. However, that information still flows through the Web, and web servers can be hacked, so there will inevitably be some privacy issues resulting from failures of security or potential use of the information by companies providing the service for advertising purposes.

This isn't a new problem. When you send an email message, you expect the contents will not be seen by other than the person to whom the email is addressed. But anything flowing through servers somewhere in the cloud could be vulnerable. Individual information that companies hold but manage to protect from hackers isn't necessarily private from governments. Google's "Transparency Report" posted November 13, 2012, said, "government surveillance is on the rise." The posting said that government demands for user data had increased steadily since Google first launched the Transparency Report in early 2010. US agencies can obtain information from Google email accounts through subpoena; there has been at least one well-publicized scandal resulting from such access and the resulting investigation. Privacy violations aren't new, but we are increasingly aware of the issue.

Web search algorithms use information on user behavior to improve the search results for others. Speech recognition can be improved by tuning to a wide variety of voices and accents. What companies generally say about this type of use of your personal data is that it isn't associated with any individual. In practice, the advantages to users of these analyses generally outweigh privacy concerns.

Information gleaned by services such as search engines is used to present supposedly relevant ads, and we all benefit from such free services. An overly restrictive privacy policy would make it necessary to charge for such services.

We are trusting companies to make significant investments in security to avoid misuse or theft of private information. Some companies may, while others will do so only after a deficiency is advertised. In particular, with hundreds of thousands of apps for smartphones, some produced by very small companies, the likelihood that there is significant effort or expense to protect the data all apps collect is small. In a well-publicized incident in September 2012, hackers obtained millions of IDs for Apple iPhones from a small company with an app for iPhones. It turned out that the company had stored them without encrypting them, a fairly simple and inexpensive process. In contrast to "too big to fail," the company apparently felt they were "too small to be hacked." The hackers in that case were apparently motivated by bragging rights rather than a malicious motive.

The company that had been hacked promised to do a full security evaluation and correct the problem. This type of episode has been repeated for both large and small companies. It suggests ironically that hackers may be motivating companies to pay more attention to security.

An indication of companies giving security a higher priority is an action by Microsoft in 2012 to drop a product that added security features to one part of its widely used enterprise software, but required a separate purchase. Security features were embedded into the core software instead, rather than being an extra-cost option. Hopefully, this trend will continue. Individuals and companies that purchase software can motivate this transition by buying products that advertise built-in security features. Perhaps a law requiring disclosure of security procedures would help for services that handled personal data, much like laws on public company reporting of financial information.

One aspect of privacy that is likely to reach public consciousness results from voice control at a distance in the home, e.g., for a Smart TV or a home control system with voice command capabilities. Some systems that allow calling across the room to select a program do the speech processing in the Internet, which means that your speech is sent digitally over the Internet. If these are always listening, then, in effect, your living room is bugged. Although the companies offering these services are unlikely to do anything but process the audio looking for commands the system can understand, there is no guarantee that these audio files will be quickly deleted, can't be hacked, and aren't susceptible to government requests. I would personally be more comfortable if (1) there was a clear indication when the device was listening, e.g., a clearly visible light on the device; or (2) the device only listened when awakened by a button press on a remote or by a "voice lock" that required saying a "wake-up" phrase that was recognized on the device itself (not in the network). For example, Microsoft's Xbox with the Kinect peripheral requires one saying "Hey, Xbox" to activate listening for more complex voice commands.

Jeff Jarvis, in *Public Parts: How Sharing in the Digital Age Improves the Way We Work and Live*, invents the useful term "publicness," describing roughly the willful sharing of information about ourselves through mechanisms such as Facebook. As the title of his book suggests, Jarvis believes that this sharing is a positive development, and recommends that we focus on the benefits rather than the downsides of letting selected audiences know more about us. He doesn't consider publicness, however, as the opposite of privacy. One shouldn't share information that isn't one's to share, he notes, for example, information about those close to us that should be their choice.

The broader picture is the tradeoff between the advantages of sharing and the potential abuse of private information. Stronger protection of individual data by companies and clear privacy policies will help avoid abuse of "publicness."

Individual and company security

I've discussed some issues of company security in the previous section. Often individual privacy and security are in the hands of companies providing services to those individuals. Companies are becoming increasingly aware of security risks to customer data and to their own trade secrets, but that awareness often doesn't translate into significant investment until there is a problem.

The trend toward Bring Your Own Device (BYOD) is complicating security within companies. IT departments often have to make business software integrate with specific devices, a development and management challenge, particularly with all the software updates for these devices and their operating systems. When a person leaves the company or loses a device, the company needs somehow to delete any confidential company information on the devices. And employees might for convenience use cloud-based storage services, leaving sensitive company data outside the company. And, since these devices are subject to the usual viruses and hacks, they may provide a route for malicious software to enter the company. Yet, the reality is that companies are better off trying to deal with these issues directly, including educating employees on the risks, rather than simply having the devices used without their knowledge or approval.

BYOD is perhaps not even the major threat to security. A 2012 survey by MeriTalk of Federal information security professionals and email administrators found that a Federal agency sends and receives 47.3 million emails daily. Despite security measures, survey respondents said that standard work email is the #1 way unauthorized data leaves their agency.

Cybercrime can be on a huge scale. For example, Global Payments, a company that processes MasterCard and Visa credit card payments, announced on March 30, 2012, that about 1.5 million credit card numbers had been stolen by hackers in an incursion that may have occurred as early as June 2011. The company indicated that it isn't sure enough data was stolen to counterfeit the cards and that cardholder names, addresses, and social security numbers were not obtained by the criminals. The disheartening aspects of this crime are that, if any organization should have heavy cybersecurity, it should be a company whose business is processing credit cards and that it took months to discover the incursion.

Part of the solution is for companies that provide the software used by other companies for common tasks such as database storage and retrieval to build in security features, as previously noted. Another piece of the solution may be accountability; if a company can at least trace the source of the intrusion, it should be able to eliminate that vulnerability. Biometric identifiers for employees logging into company systems, e.g., voice characteristics, something more secure than PINs, can help avoid intrusions. Biometric identifiers also add accountability by making it obvious who accessed a system when malicious software was introduced, so that an employee couldn't claim someone must have stolen their password when their own carelessness or bad intent was at fault.

But there is no complete solution. Like many things in our lives, we can only try to reduce risks. We won't stop using automobiles simply because there are automobile accidents, but we do try to make vehicles and driving safer with technology and laws. And we don't lock ourselves in our houses because we might be exposed to infectious germs and viruses in public places.

6

Software in Education

Education is of course a key aspect of cultural evolution. It's how we pass on the essential knowledge of our culture to a new generation. It's intended to create readiness in our children for the current world. If my assertion of the importance of language and connection with computers is correct, we are largely failing our children in the US. This has impacts on both their self-esteem and on their readiness for today's economy.

Harvard University economist Claudia Goldman was quoted in the *Wall Street Journal*, discussing the US educational system, as warning, "The wealth of nations is no longer in resources. It's no longer in physical capital. It's in human capital." Tim Cook, Apple's CEO, said in a December 2102 interview on NBC, in response to a question from Brian Williams about Apple's manufacturing in China, "Over time, there are skills that are associated with manufacturing that have left the US. Not necessarily people, but the education system stopped producing them."

In their 2004 book, *The New Division of Labor: How Computers Are Creating the Next Job Market*, Frank Levy and Richard J. Murnane discuss how the increasing incorporation of computer technology in jobs requires a change in education. They indicate that computers are enhancing productivity in some jobs, making those jobs more valuable. But, for there to be workers who can fit these jobs that partner computers and people, education must adapt to this new reality.

We must start with a good basic education. The problems in the US educational system—the dismal test scores in reading and math, among other issues—have been widely documented. Diane Ravitch is Research Professor of Education at New York University and was Assistant Secretary of Education in the administration of President

George H.W. Bush, among other influential positions. I recommend her latest book, *The Death and Life of the Great American School System*, for readers who wish a much deeper analysis than I can give of the state of our educational system and the policies that have led to an educational crisis.

It's not just an issue of students being able to read, but their comprehension of what is read. Eighth-graders averaged 265 out of 500 in vocabulary on the 2011 National Assessment of Educational Progress. Fourth-graders averaged 218 out of 500. For example, about half of eighth-graders didn't know that "permeates" means to "spread all the way through," and about half of fourth-graders didn't know that "puzzled" means "confused." The U.S. Department of Education administered the 2011 exam to a representative sample of 213,100 fourth-graders in public and private schools in all states; 168,000 eighth-graders were tested.

According to data released by the College Board in September 2012, Scholastic Aptitude Test (SAT) scores for the US high-school graduating class of 2012 fell in two of the test's three sections (math, reading, and writing), with reading dropping to the lowest level in four decades on the college-entrance test. Only 43% of the 1.66 million private- and public-school students who took the college-entrance exam posted scores showing they are prepared for college. An alternative test, the ACT college entrance exam, showed that about 75% of students *failed* to meet college-readiness standards.

And there is a crisis in the US in the growing cost of a college education. The cost of tuition in two- and four-year public colleges rose an inflation-adjusted 45% over the decade ending in the 2010-2011 school year, as an example. The increasing burden of student loans upon graduation is an increasing economic concern; it reduces disposable income and results in increasing loan defaults when students can't get jobs reflecting their educational accomplishment.

In March 2012, those who had only completed high school had an 8% unemployment rate. College graduates had a 4.2% unemployment rate. Workers with bachelor's degrees were earning 45% more in wages than those of demographically similar high-school graduates. Education, not surprisingly, makes a difference in job quality, and the distinction will accelerate in a software economy.

Even achieving a high school education seems to be a challenge in the US, one of the world's richest nations. In 2011, over twenty countries had higher high school graduation rates than the US, according to the Organization for Economic Cooperation and Development (OECD).

The US Department of Labor reported in March 2012 unemployment by level of schooling. In 2011, individuals with a Bachelor's Degree had an unemployment rate of 4.9%; those with an Associate degree, 6.8%; those with some college, but no degree, 8.7%; those with only a high school diploma, 9.4%; and those with less than a high school diploma, 14.1%.

The 2011 National Assessment of Educational Progress, an exam administered by the US Department of Education, found that only 32% of eighth-grade students were proficient in science, a slight increase over the 30% proficient in the last test in 2009. Some have argued that tests emphasizing math and reading (specified by the 2002 "No Child Left Behind" act), caused schools to de-emphasize science. The approach of measuring students by performance on periodic tests runs a risk—teachers are motivated to "teach to the test," emphasizing the parts of the curriculum that would result in higher test scores at the expense of other subjects.

In the US, the problem varies by state and region within a state, but some of the problems are largely endemic. Just to outline a few issues, based in part on sources such as Ravitch's book and based in part on limited personal research I've done in the Los Angeles Unified School District:

- *Students vary widely in ability* in a single classroom, leading to difficulty in giving all a chance to learn efficiently based on their abilities and what they bring to the classroom as a starting point. Some arrive in their first classroom speaking English as a second language. In 2010, according to the National Center for Education Statistics, some 11.8 million school-age children (children ages 5 to 17), making up 22% of the total school-age population, spoke a language other than English at home. Among them, 2.7 million (5% of the school-age population) spoke English with difficulty. Specifically, about 7% of children ages 5-9 and 4% of children ages 10-17 spoke a language other than English at home and spoke English with difficulty. And some students have had rich educational experiences through their parents at home, some haven't. Some have specific learning disabilities, such as dyslexia. And, of course, there is always a range of basic learning abilities in any classroom, with the brightest students at risk of boredom and possible misbehavior and the slowest students with simply being left behind and frustrated. In practice, teachers often must give the most attention to the lowest-performing students in the class. Students come to a classroom with more than basic ability determining their progress: Rathbun and West, in a 2004 study, found that students' race/ethnicity, poverty status, and mother's education were related to achievement.

- *In many classrooms, there are disruptive students.* This adds to the teacher's classroom management burden.

- *Too many adults in a classroom* can cause difficulty in managing the classroom. With aides for special-needs students and parents who volunteer to help, there is a prospect for students getting training at the level suited for them because of, in effect, multiple teachers in the same classroom. But, since the means of early-grade teaching is largely by speech, there is also the prospect of distraction of one group by the teaching of another.

- *Uneven skills and motivation of teachers* can leave some students who get a poor start with a difficult job to catch up in later classes. This problem has led to recommendations that teachers be ranked by how their students do, but the uneven nature of student populations in a given classroom, as well as the issue of "teaching to the test," makes this a potentially unfair and a minimally effective process. Ravitch addresses this issue in detail.

- *Focus on basic skills can shortchange learning the skills needed in a complex society.* Clearly, students need to learn reading and basic math. But focusing only on these basic skills, particularly if they are the emphasis of a test that supposedly measures classroom competence, can reduce the time spent on areas such as science and art. Higher education in math and science can result in appreciation of the beauty and excitement in these fields, but much of the education in these areas in lower grades emphasizes memorizing facts about science rather than the excitement of research and treats mathematics as if arithmetic defined the field. (Professional mathematicians speak of "elegant" proofs that have little to do with arithmetic, so there seems to be a disconnect.) Modern quantum science leads to questions about reality, consciousness, and what Einstein called "spooky" behavior that certainly belies that science must be taught as cold facts. Creative arts are also being shortchanged, an area in which unique human skills create increasing economic opportunities.

- *Distraction of non-educational programs:* Well-meaning special programs can add to a teacher's management burden. In some Los Angeles classrooms, breakfast in the classroom is part of daily activities, reducing time spent on education. Is it part of a teacher's job to encourage students to drink their milk?

As these points suggest, I believe that the issue is less the quality of individual teachers than the hurdles to teaching in the average classroom. Those classrooms present a management problem that impedes education, no matter what the teacher's skill. Teachers find themselves dealing with a wide range of abilities in the classroom, with too many students in the classroom making it essentially impossible to teach at all the levels of ability. One kindergarten teacher in the public schools (in a middle-class Los Angeles community) pointed out that some students arrive with limited English-speaking skills and no experience in reading, while others arrive reading at third-grade level because of aptitude and parental involvement, with others between these extremes. The usual result is that teachers often focus on the slowest readers, some of whom are just learning English.

Further, today's budgetary restrictions are causing movement away from smaller class sizes and are shortening the school year, making early education even less effective. We are building a long-term problem that partially accounts for the low test scores we see in later years. And, if my premise is accurate that connection with computers will be increasingly important in society and jobs, with language being a key part of that connection, the lack of language skills and computer literacy will be cause even more problems in the long term.

These basic problems of classroom management won't easily be solved. Many are a result of policies with good intent that have unintended consequences. There seems to be little study of what actually goes on in the average classroom.

And there will always be a range of abilities in a classroom. It might seem that segmenting children into groups based on their abilities would help, but doing so at the kindergarten level would be difficult and would run into parental objections when their children were put into any group other than the top-performing group. And grouping children just starting their education might incorrectly categorize a child whose only deficit was a lack of early resources at home. Such segmentation is possible at later educational levels when the students' performance has been measured for several grades, but difficult in the most critical introduction to education—kindergarten and first grade.

The role of software

A clear solution I can see for early education is to give teachers a tool that lets students progress at their own pace, measures that pace, and lets the teacher focus on helping individual students with particular problems, encouraging special talents, or recognizing advanced capabilities, with the tool aiding directly in educating the student at their natural level. Technology is already evolving today in test cases with projects such as giving students tablet computers with instructional software that adapts to their pace of learning.

The instructional software can make the teacher more effective by reporting the progress of each student. It can provide a consistent experience to each student independent of the classroom they happen to be in. It can provide a top-notch learning experience based on research and continuing improvement, just as any software improves by noting problems and fixing them. This is particularly the case if the hardware is connected and can use Web resources and the software can be automatically updated.

Software-based tools can allow a student to proceed at the pace most fitted to their skills and educational level. My personal experience in this area is as a consultant to Colvard Learning Systems, which has developed innovative early reading education software for tablet computers that uses modern speech recognition technology to allow the child's reading skill to be gauged and is optimized for the tablet computer. It was in early testing when this book was written, but the enthusiasm of students and teachers suggests that this model is highly effective.

Genevieve Shore, chief information officer and director of digital strategy for Pearson, an educational publisher, reflects a change in the way educational sources are viewed. In an October 2012 web article, she argues:

> "In education, we should no longer just provide a textbook, but a service and a similar promise to help raise achievement through course content, software and assessment. To do this, we need to invest in systems—software and resources that create and deliver engaging interactive courses; monitor use, effectiveness and engagement in new ways; and share that information with teachers and educators."

She notes that the changing nature of content is a fundamental shift that educational publishers must deal with: "This is not a freak storm that will soon give way to calmer waters."

Some schools are using educational software on PCs in the classroom or a computer lab. One example is a study conducted by Dr. Jody Woodrum, assistant superintendent of teaching and learning for grades K-5 in Bulloch County Schools in Georgia. During the 2009-2010 school year, Langston Chapel Middle School in the school system used Scientific Learning's Fast ForWord software across an entire grade. All sixth graders—including general education, at-risk, gifted and talented, English language learner, and special education students—worked on the Fast ForWord products 40 minutes a day for at least one semester. According to Woodrum, 77% of the sixth graders made English language arts gains on the Northwest Evaluation Association's Measures of Academic Progress (MAP) standard test after using the software products. For the students who achieved gains, the gains were substantial, corresponding to the 99th percentile of the Growth National Percentile Rank (GNPR), a measure of improvement relative to "academic peers" (students in similar grades and at similar achievement levels). Woodrum reported, "What was particularly interesting about the data from this study is that most students made substantial gains, regardless of whether they were below, on, or above grade level." This study suggests the power of software to allow students to proceed at the pace that best suits them.

Tablet computers are increasingly being adopted at all levels in schools, partly because they are cheaper than PCs and more portable. They can support digital versions of textbooks, which cost less than paper versions. Tablets can also support interactive teaching software, making the experience more interesting to children and adaptive to their pace.

The current generation of school age children are certainly familiar with interactive devices and will play games for hours—it would be ideal if that enthusiasm could be transferred to learning. In addition to educational publishers making their textbooks available digitally, and adding interactive elements, creative software developers could certainly create applications to teach specific aspects of subjects; for example, someone might make an animated application that demonstrated certain laws of physics visually.

Teachers, administrators, and parents can get computer reports of the progress of each student and exactly what they are having trouble with. A teacher can spend more individual time with students needing help without slowing progress of the entire classroom.

In teaching early reading, improved language technology can accelerate learning. For example, students can read out loud to a computer, and the computer can measure the accuracy and speed of their reading using speech recognition technology. A computer can ask questions orally for beginning readers to check comprehension.

Over time, children raised with computers, smartphones, and tablet computers, playing games even before they can read, might find static textbooks boring. To keep young students' attention, interactive material will be a necessity. Digital textbooks with changing content, such as science books, can be kept up to date, with online support. Breaking news and video can be part of the curriculum. We have the tools to make education exciting.

And computer technology allows a student who missed a class to catch up by going through the missed lesson, rather than just trying to move ahead despite those missed lessons. While this might seem a small problem, it isn't. Counting excused and unexcused absences, about one in 10 kindergartners nationwide is out of school for at least 10% of the year (about 18 days), making them "chronically absent," according to an estimate based on data from the Department of Education's Early Childhood Longitudinal Study.

A report published in 2008 by Columbia University's National Center for Children in Poverty found that children who missed 10% or more of their kindergarten year were the lowest-achieving group in the first grade. A 2011 Applied Survey Research study of 600 children in San Mateo and Santa Clara counties in California found that poor attendance in kindergarten and first grade may erase many of the benefits of preschool, even among those who started kindergarten with strong skills. Only 13% of children with poor attendance in kindergarten and first grade tested at grade level in reading as third-graders, compared with 77% of those with good attendance in those early grades.

The *Washington Post* quoted DC Schools Chancellor Kaya Henderson in 2012 as saying that many reformers think that "blended learning" (teachers and software) "could transform education in the United States, harnessing technology to help teachers deliver personalized lessons to every child."

Tablet computers and digital books can also make it easy for students to find and pursue material that encourages reading as they progress through higher grades. There are even lending libraries for digital books. When the choice or reading material is driven by schools, it may tend toward what are considered "classics," which, whatever their merits, may be dry reading by a young person's standards. I loved Hardy Boys mysteries in my youth (each chapter ended with something hanging, making me want to read the next chapter immediately), and my father indulged this by giving me the next book in the series whenever I finished one. (I later learned that he bought the whole series at once and hid them, giving me one at a time.) He understood that my enjoyment of mysteries would translate into enjoying reading in general. "Classics" are in the eye of the beholder; you can still buy the Hardy Boy mysteries, despite the first one having been published in 1927! And the Nancy Drew mystery series, first published in 1930, is still available.

Computer technology also lets students of similar interests form groups, irrespective of location. With today's networks, students need not be in the same classroom to share their enthusiasm.

In the US, there is a crisis in education funding. The 2008 recession hit local and state governments hard, and many cut their education budgets. In 2012, there was a political battle over a Congressional effort to raise the interest rate on student loans for college, with President Obama threatening to veto the bill if passed. This in the face of the increasing cost of college: from the school year beginning in 2002 to the school year beginning in 2012, average tuition and fees at private nonprofit four-year colleges rose by an average of 2.4% per year beyond increases in the Consumer Price Index, according to the College Board. Student debt outstanding in 2012 exceeded Americans' total credit card debt. Perhaps in part because of the cost, 43% of American students haven't completed their degree within six years of enrollment. Despite the rising costs of college tuition, most colleges find themselves strapped for funds.

In *Reinventing Higher Education: The Promise of Innovation*, the contributors suggest there is a lack of innovation in colleges. One approach being tried is heavier use of the Internet, which could, for example, allow students to "attend" a recorded lecture at a time other than it was first presented.

Teaching Computer Literacy

Reading, writing, and arithmetic must be taught, but in today's world, why not also teach computer literacy, even starting in elementary school? Simply teaching with tablet computers in basic subjects creates familiarity with using such devices, a basic form of computer literacy. Using them in early grades exposes students who don't have them at home to the devices and how to use them.

We can go much further than simple exposure. Computer literacy could be considered part of math, since many jobs using math skills today are in computer science. But well beyond math, most future jobs will involve interacting with computers. Yet, I hear little discussion of how these skills can be taught early in children's education. And why not let students in high school have the option of learning a computer language as an alternative to a human language?

My preference would be to include computer literacy as a specific topic in elementary school, giving students an introduction to computer programming at a conceptual level. To be concrete, imagine a graphical programming language that allowed generating a program by interconnecting boxes, each of which had an instruction (in natural language). Even a limited exposure to such ideas could create an understanding of what programming is. The "program" could operate within a narrow context, such as controlling the behavior of an animated character, even putting words in their mouths through text-to-speech synthesis. The result of running the program could be an animated video. A child would be motivated to "play" with such software.

A side benefit would be teaching systematic/logical thinking about accomplishing tasks, a life skill. The best programmers I've known have the analytical skills to solve problems that have nothing to do with computer programming. Good software design depends fundamentally on breaking down a problem into logical steps, a skill we all can use in areas other than software. Call the objective "logic" or "systematic thinking." I expect that readers will agree on the value of teaching children to think logically.

Computer science is clearly a key area of knowledge in today's world, and it is really part of mathematics and science, but it can get lost in early grades that emphasize the traditional areas of instruction. A key difference here is the distinction between arithmetic/algebra and "algorithms" (see the Software chapter), with the latter being very important in today's world. Regrettably, I find little discussion of these possibilities when educational reform is discussed.

Let me elaborate on the point of teaching logical thinking as a precursor to helping students eventually deal better with software as a tool and as a precursor to learning software and algorithm development. There are many logic games that one could use for older children, but how could one start early without frustrating students? I've suggested a game that resulted in animated characters doing what the instructions told them to. But let me provide another example in more detail.

Suppose the lesson involved the map of Figure 1, where the challenge is to find the shortest number of steps (tiles) to get from S (the Start) to E (the End). The shortest path has nine tiles. A key point of understanding is that the tiles are different sizes and taking paths with bigger tiles will help. Perhaps one approach would be to start with the largest block and see what the shortest route is using that block; any other path that was longer than that path need not be pursued once that number of blocks was reached, since it couldn't be the shortest path.

The point here isn't this particular puzzle or even that there is more than one way to approach a problem, but that this sort of puzzle could be discussed in early grades to help students understand systematic thinking. By then trying new puzzles, they would hopefully see the value of a systematic approach. Another point here is that teaching logic doesn't in itself require computers; the lesson could be done entirely with a whiteboard and paper handouts. Yet, thinking through a problem and understanding how it can be approached systematically is the core of computer programming and often the hardest part of writing a program.

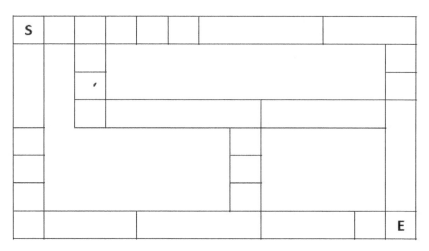

Figure 1: Puzzle: Find the shortest path by number of tiles to get from S to E.

Teaching logical and systematic thinking can be combined with other lessons. A crossword puzzle is an example of a classical puzzle that teaches vocabulary and spelling as well as logic.

I've emphasized early education because the beginning of an education can be critical to a student's later education, but later education is equally important as well. The point about computer literacy is even more relevant in high school. The point was well made by a high school student responding to the question "What is the most frustrating thing about being a high school student?" on the Quora web site in December 2012. In part, he wrote:

> ". . . this system is outdated. The job market is changing. The way we get knowledge is changing. Just last month I was forced to memorize all of Europe's countries and their capitals . . . Right in my pocket was an Android App called World Map that could have brought up all of those countries in two seconds, including their physical features. I am still appalled at how they don't teach software programming yet . . . We are not kids being prepared for a factory job anymore, we are a generation of start-ups and small businesses, a generation that is not getting the nurturing it deserves."

This section has discussed inclusion of the basic ideas of computer science and software in our education system, starting early. We've discussed basic concepts, and noted that many of these basic concepts can be taught with traditional teaching methods if teachers are given the appropriate materials and guidelines. Of course, the use of software and tools such as tablet computers can make this job easier for teachers and can allow students to move at a pace that suits their basic strengths. And teaching computer science and math on a device that is in fact a computer can make the teaching more efficient and effective, since the device itself can "execute" a student's instructions and give them immediate satisfaction.

7

Software Patents

Effective human-computer connections and user interface consistency is threatened by misuse of the patent system to patent narrow pieces of user interfaces. "Software" and "process" patents in general are often cited as an abuse of the original intent of the patent system.

The founding fathers of the United States certainly didn't envision the full role of patents today when they included in the US Constitution a clause: "Congress shall have power . . . To promote the progress of science and useful arts, by securing for limited times to authors and inventors the exclusive right to their respective writings and discoveries." There is little dispute that patent law has historically encouraged investment in research and in commercializing that research, since an individual or organization could protect itself from blatant copying of an invention. Patent law is an area of knowledge where intellectual property can be owned, and use by another can be denied or require payment to the inventor. Patents are sufficiently valuable that some firms have been formed to simply buy and license patents.

When patents are used as strategic weapons, things can get out of hand. Google and Apple's expenditures for patent lawsuits reportedly exceeded their R&D spending in 2011. Many of the patent battles over mobile phones and similar areas were over patents on software innovations.

Patent law forbids the patenting of "mathematics," supposedly because it is a "law of nature," a "scientific truth," and, as such, it can never be "invented," only "discovered," and patents are not granted for discoveries. This can get a little fuzzy with certain areas such as drug development involving a great deal of discovery of scientific truth.

What is considered math today is narrowly interpreted, since patents have been granted on mathematical algorithms implemented in practice as software. Excluding software patents entirely would lead to definitional issues—most inventions today would have software in any practical implementation. The issue is typically addressed in patent applications by emphasizing that the invention can be built as a machine implemented on computer hardware, making it a machine instead of a mathematical process. At the time the first patent law was written, mathematics had only an abstract existence. One could use math to show that certain assumptions led to certain conclusions, a "discovery" by derivation rather than an invention (perhaps). But math is intrinsic in engineering; we'd be uncomfortable driving a car that was built with no math involved in its design. Patent law has thus simply adjusted to the realities of modern technology.

Nevertheless, the fact that software can execute any process in a digital implementation can lead to abuse. Software patents are often some of the most controversial. Some seem "obvious" (intuitive) rather than a true invention; a patent application can be denied on the basis of being obvious. But what is "obvious" is certainly subjective, and the patent office struggles with denying patents on this basis.

Another basis for denying a patent is if it has been done before, and the fact that it has been done before is documented, constituting "prior art." The increasing pace of technology and its volume can make it difficult for a patent examiner to find prior art if it exists.

The America Invents Act, an attempt at patent reform passed in 2011, makes the date of patent application, rather than the use of the invention, define priority, changing US law to match most of the world's approach. One unfortunate result is that the Act motivates filing anything that might be patentable as quickly as possible.

Business process patents are similar to software patents. For example, Amazon has a patent on its "one-click" buying option, a business process implemented in software.

Making a business of patents

Nathan Myhrvold, the former Chief Technology Officer at Microsoft who founded Intellectual Ventures (which owns tens of thousands of patents and makes money by licensing them rather than using them to make products), has complained about the "culture of intentionally infringing patents" in the software industry. Myhrvold told *Business Week* in 2006: "You have a set of people who are used to getting something for free."

But there is a distinction between a company that is in a business suing a competitor it feels is using its invention illegally, and a "Non-Practicing Entity" (NPE), sometimes called a "patent troll," that sues with the intent of getting damages or a license fee, but doesn't use the patent itself. An important part of the distinction is that a practicing entity can be counter-sued for using the patents of the company being sued; historically, this usually leads to some sort of cross-licensing settlement, although, recently, such battles have become more strategic and more difficult to resolve (more on this later). NPEs are growing as organizations devoted to collecting patents and deriving revenues from them are growing larger. The *Wall Street Journal* reported in July 2012 that one such entity had revenues of over $100 million in the first quarter of 2012, and that many practicing companies are trying to make their patent portfolios generate more revenue, in some cases by placing them in a separate entity dedicated to collecting license fees. According to a Boston University study, NPE suits cost American technology companies over $29 billion in 2011.

Damages in patent suits reflect the difference between an NPE and a practicing entity; an NPE may not be able to assert that the patent infringement damaged it as much as a practicing entity, the patent of which was violated by a competitor. Often a company sued by an NPE will settle for less than the cost of litigation, even if they feel they would prevail in court, and the NPE may be motivated to accept smaller returns from each company sued and make it up in volume. With this practice, the patent's validity isn't challenged in court.

The growing burden of patent suits

The growth of patent suits can lead to companies spending significant resources to file defensive patents and license others; small companies would be at a disadvantage in this environment.

Although it is difficult to get individuals to talk about it on the record, principals of a number of small companies that were acquired have indicated privately to this analyst that they received a lower price than they felt they deserved for their company because a larger acquirer used threats of patent suits to force the acquisition and discourage counterbids.

In late 2012, Cisco even went to the extent of formally accusing an NPE of "racketeering," extorting payments from customers of its products by the threat of large legal fees defending against patent suits for Cisco products. Cisco claimed the suits were filed without a valid reason for such threats, and were simply extortion.

A bipartisan bill in Congress is intended to address the burden of patent suits by NPEs on startups. The "Saving High-tech Innovators from Egregious Legal Disputes" (SHIELD) Act was introduced on August 1, 2012, by Democrat Peter DeFazio and Republican Jason Chaffetz. The congressmen said the act was intended to save jobs at technology start-ups. DeFazio said in a statement: "Patent trolls don't create new technology and they don't create American jobs. They pad their pockets by buying patents on products they didn't create and then suing the innovators who did the hard work and created the product . . . My legislation would force patent trolls to take financial responsibility for their frivolous lawsuits." The wording of the bill targets any plaintiff "who did not have a reasonable likelihood of succeeding" in their patent lawsuit. "A single lawsuit, which may easily cost over $1 million if it goes to trial, can spell the end of a tech start-up and the jobs that it could have created," Chaffetz said in a statement.

The proposed Act singles out and attempts to define software patents. According to the Act, a software patent "covers any process that (A) could be implemented in a computer regardless of whether a computer is specifically mentioned in the patent; or . . . (B) any computer system that is programmed to perform a process described in A." Given that most products of any complexity today include a digital processor (or could), the definition of "computer" could become an issue. The term "process" might also be difficult to limit. A washing machine has cycles, usually controlled by software, and a washing machine implements a process—wash, rinse, spin-dry. Is a patent on a washing machine a software patent? Despite definition issues, the remedy suggested by the Act—creating a penalty for essentially frivolous patent lawsuits—may discourage such lawsuits, and seemingly should apply to all patents. A frivolous lawsuit on any patent is against the public interest. A company that is penalized for a few frivolous lawsuits would presumably be more circumspect in the future.

Some industry leaders clearly see the issue. Amazon CEO Jeff Bezos was quoted by the Metro news site in October 2012 as saying that innovation and society itself was threatened by the patent lawsuit culture. Calling for new legislation to be introduced by national governments, Bezos reportedly said, "Patents are supposed to encourage innovation, and we're starting to be in a world where they might start to stifle innovation. Governments may need to look at the patent system and see if those laws need to be modified because I don't think some of these battles are healthy for society."

Consumers are often battlefield casualties. Some of the patent battles that are emerging, for example, could affect the usability of your smartphone. Several Apple patent suits against mobile phone manufacturers, for example, are widely regarded as an attack on Google's Android operating system. Apple's well-publicized triumph over Samsung phones using the Android operating system in August 2012 is an example of the seriousness of the issue. The emotional aspect of this battle is in part reflected in Walter Isaacson's biography of Apple's Steve Jobs, where Jobs is quoted as feeling that Google in effect stole the iPhone design, and he planned to fight hard to rectify that injustice. (Jobs supposedly used less polite terms.)

The conflicting issues are perhaps summarized in statements issued by Apple and Samsung after the August verdict. Samsung said in a statement: "Today's verdict should not be viewed as a win for Apple, but as a loss for the American consumer. It will lead to fewer choices, less innovation, and potentially higher prices." Apple's CEO Tim Cook said in an email to employees: "We value originality and innovation and pour our lives into making the best products on Earth. And we do this to delight our customers, not for competitors to flagrantly copy." On this subject, *The Economist* news magazine found an interesting quote from Apple's Steve Jobs from a television documentary in 1996: "It comes down to trying to expose yourself to the best things that humans have done—and then try to bring those things in to what you're doing, and we have always been shameless about stealing great ideas." [The reference may have been in part to the use by both Apple and Microsoft in their operating systems of the Graphical User Interface, ideas developed at Xerox's Palo Alto Research Center.]

Some of the Apple patents relate to the user interface on the phones. A selection of the cited patent titles in an Apple suit against HTC illustrate the user interface content: "Touch Screen Device, Method, And Graphical User Interface For Determining Commands By Applying Heuristics," "Unlocking A Device By Performing Gestures On An Unlock Image," "List Scrolling And Document Translation, Scaling, And Rotation On A Touch-Screen Display," "Automated Response To And Sensing Of User Activity In Portable Devices," and "Message Protocol for Controlling a User Interface From an Inactive Application Program." Google patents provided to HTC for a countersuit against Apple include some things that would apparently limit Apple's user experience, including ways to upgrade software wirelessly and store user preferences. Whatever the merits in this specific battle, it is only one example of patent battles that attempt to claim ownership of certain aspects of how we might interact with our mobile devices. If fully enforced, as opposed to cross-licensing and/or reasonable fees for use of the patents, certain aspects of user interaction could be more clumsy than necessary because of patent workarounds.

In an entry on the Becker-Posner Blog posted September 30, 2012, Judge Richard Posner (the well-known jurist who sits on the 7th US Circuit Court of Appeals in Chicago, teaches at the University of Chicago, and has written books on intellectual property and the impact of law on economics) suggested that patent and copyright law can restrict competition and creativity excessively. Previously, Posner presided over a 2012 Apple lawsuit against Motorola Mobility, now part of Google. Posner rejected Apple's request for an injunction barring the sale of Motorola products claimed to be using Apple's patented technology. In his ruling, Posner said an injunction barring the sale of Motorola phones would harm consumers. He further rejected the idea of trying to ban an entire phone based on patents that cover individual features like the smooth operation of streaming video. Apple's patent, Posner wrote, "is not a claim to a monopoly of streaming video!"

In the September 2012 blog entry, Posner said he is concerned that patent protection may be excessive. He said that the ratio of the cost of inventing to the cost of making copies of the invention should be considered in patent protection, giving more protection for inventions such as drugs where the cost of invention and approval for sale is high, but it takes many sales (because of the low cost of making copies of the drugs) to recover the investment. He notes that for products where the ratio is low, the firm with the invention generally receives benefits from first entry, affording the firm with the ability to improve the invention ahead of competitors, and identification with the invention. That first-mover advantage can provide a long-term advantage well after a shorter-term patent expires.

Posner wrote that, when patent protection provides an inventor with more insulation from competition than needed to have an adequate incentive to make the invention, there are negative results, including increasing market prices above efficient levels, engendering "wasteful patent races," and encouraging patent "trolls." Posner singled out the software industry as an example of "the problem of excessive patent protection." He said that "the conditions that make patent protection essential in the pharmaceutical industry are absent." He noted that today's digital devices such as cellphones, laptops, and tablets can contain "hundreds of thousands of separate components (bits of software code or bits of hardware)," all of which could be separately patentable, creating "huge patent thickets." Posner noted the difficulty for patent examiners, judges, and juries to properly understand and evaluate claimed software inventions. In a *New York Times* article on October 7, 2012, Posner is quoted as summarizing simply, "The standards for granting patents are too loose."

As this book was completed, the courts were increasingly taking a position that sales of a product such as a smartphone with many features could still be sold even if it violated a patent on a feature of the phone. The courts seemed to be shifting the burden of proving harm by sales of the infringing device to the suing party, rather than the assumption that use of an infringing patent was enough to indicate harm. By doing so, the impact of software patents on aspects of a product not necessarily fundamental to the product's use is reduced. This might encourage licensing the patent on reasonable terms.

Too many patents?

With so many patents having been issued and new ones being issued daily, and with the patent office taking years to approve patent applications (there are over a million patents pending at the time of this writing), it isn't economically feasible for a firm (particularly a startup) to research whether a new product may violate some patent that might cover one of many features. Since the meaning of claims in a patent depends on the text in the body of the patent, even the analysis of one patent can be time-consuming. Some patents are invalidated in court, so a complete evaluation would include finding prior art that might invalidate the patent.

There are exceptions; if the company producing the new product is using a major innovation by a competitor, and there is a strong likelihood of a patent on the major innovation, there can be focused research. Given the complexity and number of patents, however, in most cases, companies launching products must simply take the risk of being sued and hope they can negotiate reasonable terms or a patent trade if there is a valid claim of infringement.

The increasing litigation in patents encourages the filing of patents on almost any innovation as a defensive measure. The number of patent *applications* filed with the US Patent and Trademark Office rose to over 535,000 in 2011, from 165,000 in 2000; the number of *patents granted* rose to 247,713 in 2011 from 175,979 in 2000. This explosion of patents could easily stifle innovation, harming the purpose of the patent system to encourage innovation.

In the extreme, one might assume patents could be forbidden for software. This may be infeasible since so many products are implemented with digital technology (and supporting software) today. Trying to exclude a category such as "user interface patents" might also be difficult, both in terms of defining the category and passing laws to change the patent system. Both approaches would probably have to leave patents already issued (and perhaps existing patent applications) in force, thus having little near-term impact.

Partly because of the difficulty of understanding what is truly a new invention in software, some granted patents don't seem to rise to the level of innovation that goes beyond "obviousness." An Apple slide-to-unlock feature for a mobile phone is patented (Patent #8,046,721). A judge in the UK in July, in the Apple suit against HTC, called that patent and two other user-interface patents (one on touching with more than one finger at a time and the other on a multilingual keyboard) invalid. The judge said the slide-to-unlock was an obvious development, citing the presence of a similar feature on a 2004 Swedish phone. While HTC could afford to find this obscure prior art to present to the judge, one wonders how the patent office could do so without expanding its resources significantly.

Apple has sued based on design patents as well. It sued Samsung on its Galaxy tablets, which have the same look as iPads from the front (a glass panel with a border). A judge in the UK ruled against Apple in July 2012, saying that the Samsung design differed enough overall so as to not infringe Apple's intellectual property. Samsung said in a statement, "Should Apple continue to make excessive legal claims in other countries based on such generic designs, innovation in the industry could be harmed and consumer choice unduly limited."

Patents on the user interface

Is such ownership of user interface and general design ideas in the public interest? Will it result in crippled devices if certain powerful ideas (and some might say, intuitively obvious ideas) are reserved for one company?

History may provide insight into this issue. Looking at the evolution of the interface between people and machines, there have been a series of de facto standards that accelerated the use of technology. Before what we think of now as "user interfaces," there were buttons and switches. Every piece of electronic and computer equipment has them. The most fundamental are on-off switches, but then there are remote controls (with too many buttons, perhaps). When you put a lot of buttons in an array and label them with letters, numbers, and symbols, you get a keyboard, clearly a user-interface option. Imagine if only one company could use push buttons for two decades. The idea isn't so far-fetched if one looks at the nature of recent patents.

And what about personal computers? The WIMP interface (Windows, Icons, Menus, and Pointing device) led to a revolution in usability and popularity of personal computers, and has to a large degree been adopted in varying forms for smartphones and pad computers. At this point, we understand how to use a new device at a basic level using this paradigm without recourse to a user manual. It's not tough to understand that the operation is the same whether the pointing device is a mouse or a finger, for example. There are many aspects of the familiar Graphical User Interface (GUI) paradigm on PCs beyond the basic WIMP interface, e.g., double-clicking to open an application or file. What if every vendor had to come up with a new way to open a file? What if one company had a monopoly on one aspect of the WIMP interface and another on a different aspect of the WIMP interface and refused cross-licensing? Utility to users and the growth of technology would clearly be hampered. It is hard to believe that what amounts to a limitation on innovation was the intention of the US founding fathers when they required the development of a patent system in the Constitution.

And what about web search? The universal search box pioneered by Google is certainly a user interface innovation. In this case, the technology is less visible, with the real innovation in the network-based processing and in the software analyzing the Web continuously to be able to produce immediate results. But, whatever analysis is behind it, when you see a search box on a web page or in a browser, you know how to use it, and you know what to expect as a result. This category of user interface—where the invention is largely network-based processing—is increasingly important, and is also part of what we expect to be consistent as users. In this type of user interface, the key invention is the operation of software within the network (the "cloud"). To a large degree, this more "invisible" invention is protected by trade secrets, since it isn't visible on the screen. Processing to assess the relevance of Web pages to search terms relies on methods and data that can (at least in theory) be protected by patents. Patents in this area can certainly be granted, but filing patents for such technology has the disadvantage of requiring disclosure of the methodology, and enforcing such patents is difficult when the method of operation is hard to discern from outside observation. Patents on such technology may not serve either the company's interests or the public's interest. But, if they also become part of the current patent "game," there are even more questions about their being in the public interest.

The basic user-experience technologies discussed are at the heart of our everyday use of technology. It is in the public interest to have these be at least *de facto* unifying standards, both in terms of making technology easier to use and avoiding monopoly pricing of technology by a firm that "owns" an effective aspect of user interface design. In *Trillions*, Lucas, Ballay, and McManus point out that a single company can create parts of a pervasive system (with parts such as pad computers, laptops, and smartphones communicating) that work only for that brand, creating a requirement that a buyer of one part stick with that brand when buying another part to get full functionality. The authors call this "the challenge of pervasive computing." They ask, "How do we build a world that hangs together well enough to be worth living in without giving up the ability to pick and choose among competing offerings?"

Perhaps industry groups could encourage cooperation on standards, and gain cooperation as companies realize that the patent wars are a lose-lose battle. Historically, many formal industry standards have involved patent licensing. The creation of standards that clarify core user-interface elements may be one approach to allowing consistent user interfaces across brands. Many Standard Setting Bodies (SSBs) require the parties involved in the standard-setting process to disclose information regarding relevant patents in order to include the relevant information in the standard-setting process. If any relevant patent (or patent application) exists, many SSBs require the patentee to agree on specific licensing conditions. The conditions may be that the license must be granted under Reasonable And Non-Discriminatory terms (RAND licenses) or that the license must be royalty-free. In the Apple-Samsung suit mentioned earlier, Samsung counter-sued with patents involved in a standard (unsuccessfully).

One way to address the situation where different patentees own a number of patents relevant to a user-interface paradigm is to create an open standard and set up a patent pool that eases the process of licensing patents required for the standard. This solution isn't a quick fix (even if companies were willing to cooperate to create it), since defining a standard is a slow process.

Posner's point about the huge number of interacting software and hardware components in a digital system driven by software makes the issue even more complex. Should a device or software program or service be removed from the market because one minor element is found to be infringing? Judges can decide that awarding damages is sufficient, rather than taking a product off the market as a remedy. "Judicial restraint" that limits injunctions and awards could make the launching of patent suits less attractive. As this is written, appeals courts seem to be supporting this point of view.

Beyond the question of whether patents on aspects of user interaction are in the public interest, it has become increasingly difficult to judge when something is a true invention or just an embodiment of something obvious being implemented in software. The issue is particularly difficult with some of the user interface innovations being applied with mobile devices such as smartphones and pad computers. We might express that something is bigger by spreading our fingers in a familiar gesture; is it then patentable if one applies that gesture to a touch screen to enlarge the image? Since user interface options are most effective when they are intuitive (we knew how to point before we had PCs, for example), the issues of obviousness are particularly subtle with user interface innovations.

The patents in some cases skirt obviousness by suggesting specific methods for deciding what constitutes an action such as spreading the fingers (as opposed to accidental touches). The patent "claims" (indicating what the patent specifically covers) may thus be much narrower than the description of the patent would suggest. The battles in patent court thus can become quite complex (and expensive) as one company accuses another of infringement and the accused refutes by saying in effect "we don't do it that way." Certainly, practicing patent law is currently a growth industry. Some patent firms are hiring PhDs to help lawyers to analyze patents and even to help draft them (perhaps with the advantage of creating a new class of jobs).

Helping the patent office

The challenge to the patent office is clear. Each application isn't necessarily reviewed just once. An illustration of how a large firm can wear down the patent office is perhaps illustrated by the history of Apple patent #8,086,604 ("A universal interface for retrieval of information in a computer system"), which could be considered to cover aspects of Apple's Siri personal assistant, as reported in an article in the *New York Times* published in October 2012. The patent application was originally filed in 2004. The application (or at least parts of it) was rejected by the patent office eight times, but Apple changed wording of the patent each time or otherwise argued against the rejection until the patent was granted in December 2011, seven years after submission. One point here is that the patent may not be as broad as its title and abstract would suggest, since this process normally narrows what the actual claims cover.

The most cost-effective time to review a patent application in depth is during the review process, not after the patent is issued and must be challenged. The patent office is testing a "peer-review" process in which experts in a given technology can submit what they believe is prior art when a patent application is being examined. Participation in the experimental peer-review process is currently voluntary, at the discretion of the patent applicant, with the benefit largely being faster processing of the patent application. It seems unlikely that an applicant with a marginal patent would submit to peer review, so this may have minimal impact on borderline patents. It would apparently require a revision in the law to compel submission to such an outside review. But the patent office by itself is overwhelmed; less than 8,000 examiners received more than half a million applications in 2011 alone.

The recently passed Leahy-Smith America Invents Act, an attempt at patent reform previously noted, largely targets reducing litigation and patent office costs by narrowing one basis for challenging patents, e.g., by giving patent rights on a first-to-file basis, rather than first-to-invent, as previously noted. While it may limit one aspect of patent challenge, most analysts feel the first-to-file feature will create an incentive for filing almost any idea as quickly as possible so that someone else doesn't beat the filer to the patent office.

The US Patent Office publishes patent applications eighteen months after the effective filing date of the application (before approval of the patent) based on a statutory mandate contained in the American Inventors Protection Act of 1999, and the Act makes outside review possible before the patent is granted. If industry organizations reviewed every patent application in their field through crowd-sourcing (essentially a blog where individuals or companies can note prior art in response to a posted patent application), they would be unofficially helping patent agents. The presence of such a practice might discourage the filing of questionable patents (or at least increase the cost of doing so). Companies would be motivated to contribute to the "blog" through reviewing the submissions of their competitors. Since companies can review patent applications before they are granted under the recent law change, this shouldn't create any more impediments to valid patents being granted than is currently the case, but it might make it easier for individuals with knowledge of prior art to participate.

Company cooperation on the decline

To the degree that such patents existed in the past, most companies had a licensing policy that didn't forbid use of their patented technology, but required licensing fees or cross-licensing agreements. Largely, companies typically didn't attempt to use the patents to give their products an exclusive advantage. For example, Qualcomm claimed patents that covered technology used in the CDMA wireless transmission standard, but reached a legal settlement in 1999 that allowed broad use of the standard, and continues to collect licensing fees.

In September 2011, Microsoft seems to have taken a similar settlement approach in an announced agreement with Samsung on the use of Microsoft patents in Samsung's Android-based phones. Samsung will be paying Microsoft a royalty on every Android phone, in effect placing a price on the "free" Android operating system; Microsoft's Windows Mobile operating system, which carries a licensing fee, is thus on more even ground. Microsoft didn't forbid sale of the Android system or attempt to cripple it, but put a price on it. This approach avoids limiting user interface technology, but it also means that every manufacturer has to reach a separate deal, perhaps with multiple companies, and the licensing fees are likely to be reflected in higher prices to consumers.

The status quo is that full user interfaces for smartphones largely exist as islands defined by mobile phone operating systems. The OSs that dominate the future of smartphones today come from Apple (iOS), Google (Android), and Microsoft (Windows Mobile), with Research in Motion planning to launch another alterative at this writing. The importance of an effective user interface is clear—Apple proved with the launch of the iPhone that ease of use could drive smartphone sales. If the OSs shared the most valuable user interface features, the interests of the public in being able to move from one to another with minimal difficulty would be served; but patent fights threaten even the consistency that exists today. A company with a superior device deserves a premium, but not the premium they would get if they were so unique as to lock in a user who had invested in learning the platform. With the deep pockets and huge patent portfolios of the companies delivering these OSs (now that Google has expanded its patent stash with the Motorola Mobility acquisition and purchases of IBM patents), there is a prospect of protracted legal battles that aren't in the interest of the consumer and use company cash that could be otherwise spent on creating new innovations.

Possible solutions to the patent problem

Is there some change that would better support the original objective of the patent system? The earlier discussion of a law that penalized "frivolous" patent suits could be part of the solution. Industry cooperation is another, but unlikely.

Perhaps further patent reform could encourage more cooperative behavior by setting a standard (low) patent annual licensing fee that is significantly less than the cost of going to court. If a company that is asked to license a patent offers to pay the patent owner that statutory fee, the law could place a burden on the patent owner that refused to accept that offer. If the patent owner, despite the offer of the standard fee, sued for more, they would first have the burden of proving that the fee didn't cover the value of the patent before getting into other issues. If the patent covered one feature among many in a product, it might be difficult to prove extensive damages. As part of the law, the fact that a company was willing to pay a minimal license fee to avoid legal expenses could not be used as evidence that they were in fact infringing. This approach might have the effect of making it easier for patent holders to be compensated for a valid invention without having to search for payments that cover high legal expenses, making the "tax" of the patent system on companies more acceptable.

A longer-term approach to reducing the problem is simply to reduce the term for which patents are granted. When the current 17-20 year term was originally made law, it might have seemed that two decades was required to get full value for an invention. The acceleration of development and deployment in our digital economy would seem to justify reducing that period. If the term were reduced to five years (perhaps gradually over time), it would seem sufficient to give a firm with a strong patent enough time to establish its product and brand as the leader, with continuing innovations helping to preserve the lead.

A new law could allow classifying patents and assigning patent terms based on the investment required to create the patent, the approach suggested by Judge Posner; thus a drug patent might get a longer term than a user interface patent. The former takes years and much investment to validate the drug; the latter doesn't usually require an expensive research effort. Another alternative would be to amend patent law to require that patents more than three years old should be licensed at "reasonable and non-discriminatory terms," with a maximum royalty, unless the owner can show that a significant investment in the technology has yet to be recouped.

If the term of patents is reduced, patent law could allow a procedure for extending a patent term beyond a much lower standard term if substantial investment or delays were required before the patent could be turned into a product. Perhaps a measure such as Posner's ratio of the cost of invention to the cost of producing the product could be employed, with a large fee required for initiation of such a review by the patent office. Many patents would thus be void after the lower standard term, reducing the costs and complexity of patent issues.

The public interest would be served by shorter basic patent terms in two ways: (1) there would eventually be fewer patents in effect if there is a shorter patent term, reducing the burden of patent litigation on the economy; and (2) patented ideas would become available for use in a time frame that could allow the best to be used uniformly across products and services, of particular importance in the user interface. Such patent reform, if it doesn't affect existing patents and applications and gradually reduces patent terms, might be politically feasible. This is unfortunately not a short-term remedy, but nevertheless one that can eventually reduce a critical problem.

Passing laws or setting up prior-art reviews aren't short-term solutions to the existing patent thicket. One short-term solution is judicial restraint, one option that courts seem to be adopting, as previously noted. If a company is enjoined from selling a product because of an incremental feature, the suing company has won, even if the suit fails at a later point. Even if the feature can be removed in an updated version, the loss of time on the market can be deadly in today's rapidly changing product environment. And how would you feel as a consumer if a company removed a feature in a device you already bought as part of a "software update"?

8

War and Cyberwar

Another area affected by the increasing role of software in society is the vulnerability of developed countries to attacks on that software—"cyberwarfare." In conventional warfare, technology has revolutionized modern military and intelligence organizations.

This section will touch on only a few trends where software is having the most impact on warfare, and what those trends may mean for the future of society. A concern is that, while we are well prepared and spend significant funds on war involving soldiers and weapons, we are less prepared for cyberwar and more susceptible to its affects relative to less-developed countries.

War

We may be approaching a significant change in human society in the ways that conflicts are resolved between nations or between nations and organizations that wish to change them (e.g., terrorist groups). Wars have historically been a matter of large numbers of soldiers and weapons manned by those soldiers encountering a hostile group of similarly equipped soldiers.

Because of the tighter integration of economies between nations with large militaries, conventional wars between those nations seem unlikely. There should be a natural reluctance to going to war with your customers or creditors. Some smaller nations are motivated to create an external "enemy" to unify their populations, however. For those cases, technology is increasingly important in managing the threat. Good intelligence information and minimal responses such as economic sanctions can moderate the threat.

Small groups have used the disruptive effects of terrorist acts to promote their agenda, and this is increasingly a challenge faced by both large and small nations. The key to the effectiveness of terrorist acts is that it is easier to attack than defend, particularly when the attacks are from individuals or small groups that can hide in the general population. Attempting to prevent all such attacks is almost impossible, just as preventing all crime is impossible, but intelligence and limited responses such as drone attacks may be a balanced approach. Technology and human agents used to gather intelligence might be a key weapon against terrorism at this level.

Software has long been part of the evolution of weapons, and is becoming more important as calibrated responses become more prevalent than conventional warfare. Much of the contribution of software is in the details of how weapons systems work; e.g., stealth aircraft are unstable because of the radar-reflection-reducing geometry they use, and require computer aid for stable flight. Drones, with abilities to communicate and be self-directing up to a point, are an example of how technology can help conduct limited campaigns against smaller groups. In intelligence agencies, software is necessary to collect and analyze the huge volume of data that can be obtained from communications, satellite systems, agents, and other sources. As manager of the Computer Science Division of a small aerospace company early in my career, I experienced the willingness of the Defense Department and intelligence agencies to examine advanced ideas and use those that work. We often hear about the problems with large, hard-to-manage weapons systems, but there are many smaller efforts that one doesn't hear about (sometimes because they are classified) that are very effective. There are heroes in the civilian workforce of the Defense Department and intelligence agencies whose contribution to national security is unlikely to ever be disclosed.

Conventional warfare or small terrorist actions are unlikely to be the major issues of the next decade. Instead, war by software—cyberwarfare by other countries or terrorists—is the major threat that countries with well-developed economies such as the US will face.

Cyberwarfare

Cyberwarfare that attacks computers and the systems they control, when conducted by governments or terrorist organizations, is particularly difficult to deal with. The potential for cyberwarfare on a large scale has deep implications for national security, and is more dangerous than most people want to recognize.

Cyberwarfare is most likely occurring today at what amounts to a "probing" level, trial attempts to see what can be done without retribution. In October 2012, several financial institutions experienced a "denial-of-service" attack on their web sites that made it difficult for their customers to use those sites, and news reports of the attack indicated that "sources" believed it was sponsored by Iran. A country might feel it can more easily get away with a cyber attack—perform the attack anonymously—than with conventional warfare.

Denial-of-service attacks have a minimal long-term impact at the level they have been employed to date. There are much more destructive attacks possible, attacks that destroy data or even cause failures in electrical generation and distribution systems and other critical infrastructure. In his book, *Cyber Attacks: Protecting National Infrastructure*, Edward G. Amoroso, Chief Security Officer for AT&T, commented: "The current risk of catastrophic cyber attack to national infrastructure must be viewed as extremely high, by any realistic measure. Taking little or no action to reduce this risk would be a foolish national decision."

We know that the impact of cyberattacks can be significant. In 2010, Iran's nuclear development was reportedly hit by a virus from unknown sources that was targeted at centrifuges refining uranium, supposedly causing some destruction and setting back progress.

In a video obtained by the FBI in 2011 and released to the public in 2012 by the US Senate Committee on Homeland Security and Governmental Affairs, an al Qaeda operative calls for "electronic jihad" against the United States. The al Qaeda video calls for cyberattacks against the networks of the government and critical infrastructure such as the electric grid. The Homeland Security Committee reported the Department of Homeland Security received more than 50,000 reports of cyber intrusions or attempted intrusions between October 2011 and May 2012. In the last of the 2012 Presidential campaign debates, President Obama said that new areas such as cyberwarfare must receive increased funding as defense priorities shift.

In October 2009, a general took charge of the new US Cyber Command, a military organization charged with using information technology and the Internet as a weapon (and presumably defending us against cyberattacks). Similar commands exist in Russia, China, and other nations. The senior intelligence official at the US Cyber Command, Rear Adm. Samuel Cox, has said al Qaeda operatives could purchase the capabilities for cyberattacks from criminal hackers. The need for a new Command arises in part because the agency with the most knowledge of the cyber world in the US is the National Security Agency, and its operations are legally confined to gathering intelligence rather than either defending against cyberattacks or counterattacking.

In October 2012, US Defense Secretary Leon Panetta stated in a speech that cyberthreats had grown, and the Pentagon is prepared to take action if America is threatened by a computer-based assault. He was reacting in part to cyberattacks on Persian Gulf oil and gas companies, attacks suspected to have been sponsored by Iran. In his speech, Panetta said a virus replaced crucial system files at Saudi Arabian state oil company Aramco with the image of a burning US flag, and also overwrote all data, rendering more than 30,000 computers temporarily useless. Panetta said an attack on Qatari natural gas producer RasGas was similar. He indicated that the Pentagon has invested billions in improving its ability to identify the origin of a cyberattacks, block them, and respond when needed.

In the same month, in another talk, Department of Homeland Security Secretary Janet Napolitano said, while recent news had been about financial institutions being hit with Distributed Denial of Service attacks, the nation's control systems for major infrastructure, such as utilities and transportation infrastructure, were also being targeted. Napolitano used the example of the economic cost of tropical storm Sandy on the East Coast as an example of the damage caused by even temporary electrical outages.

In November 2012, preceding defeat in the Senate of the Cybersecurity Act introduced by Senators Joe Lieberman and Susan Collins, Senate Majority Leader Harry Reid said in a statement, "National security experts say there is no issue facing this nation more pressing than the threat of a cyber attack on our critical infrastructure. Terrorists bent on harming the United States could all too easily devastate our power grid, our banking system, or our nuclear plants."

Other US authorities have warned that foreign Internet hackers are probing US critical infrastructure networks, including those that control utility plants, transportation systems, and financial networks. "We know of specific instances where intruders have successfully gained access to these control systems," Panetta said in his October 2012 speech. "We also know that they are seeking to create advanced tools to attack these systems and cause panic and destruction, and even the loss of life." Panetta said the Defense Department is investing more than $3 billion a year in cybersecurity to beef up its ability to defend against and counter cyberthreats, including investment in the US Cyber Command. Emphasizing the potential for cyberthreats and the need to prevent such attacks, Panetta said that the US was facing the possibility of a "cyber-Pearl Harbor" and emphasized, "This is a pre-9/11 moment."

Cyberwarfare can't be fully understood without a broader context. Terrorism draws part of its effectiveness from being inexpensive warfare that small groups or countries can easily conduct. Terrorist groups will often claim responsibility for an incident when they believe it is in their interest, but, even when they do so, the actual connection may not be clear and is often impossible to prove. The psychological impact of conventional terrorist attacks such as a suicide bombing often is much larger than the actual damage. Computer terrorism doesn't require the perpetuators to give up their lives or to even clearly identify themselves, yet could cause much more damage.

Even when the source of a cyber attack is identified, it can be difficult to respond with an effective response in kind. A key aspect of cyberwar is that, the more developed a country and thus the greater use of computer technology, the more vulnerable the country. The US may be the most vulnerable country to cyberwarfare because of the key role of software in our infrastructure and businesses. In conventional warfare, superior weapons and the size of a military is the greatest determinant of the outcome of a conflict. In cyberwarfare, the vulnerability of an attacker of the US to a cyberwarfare counterattack may be small. If their industry and government makes little use of software, e.g., is largely agricultural, it may be difficult to do much damage with a cyberattack.

Another factor is the relative openness of networks in the US and the wide use of portable and connected digital devices such as smartphones, tablet computers, and laptops. Many viruses are introduced into companies through the personal digital devices of individuals connecting with internal company networks, as I noted earlier in discussing the Bring Your Own Device issue, making defense of company software more difficult. Countries with more control over their Internet connections with the rest of the world can more easily block a cyberattack over the Internet by activating a firewall.

The imbalance of US vulnerability versus some potential attackers means that it will be more difficult and expensive to mount an effective defense for the US than some other countries. Our sophistication in software is part of our means of defense, but also part of our vulnerability. Richard A. Clarke (the first Special Advisor to the President for Cyber Security in 2001) in his book with Robert Knake, *Cyber War: The Next Threat to National Security and What to Do About It*, put it simply: ". . . cyber war places this country at greater jeopardy than it does any other nation."

Another typical attack over the Web, theft of intellectual property from a company, e.g., software or trade secrets is harmful to the economy, but largely gets shrugs from citizens outside the company, who don't see it as directly impacting them. Nevertheless, the impact to the economy may be much more substantial than the more obvious denial-of-service attacks.

Most analysts believe that countries, if not terrorists, are capable of much more serious attacks, e.g., attacks that would disable a country's defense systems or steal secrets of weapons systems. Many of these could attack certain aspects of conventional weapon systems; a warship has a complex computer system that, once interfered with, could make it hard for the ship to be controlled, for example.

Cyberwarfare has the potential to go beyond economic damage, to take lives. An attack on the software running an electric grid would be dangerous. If a nuclear plant's controls were successfully targeted, the results could be disastrous.

Secure internal networks supposedly isolated from the Internet are no more secure than the people using them. One internal spy, given the right software to load, can set those systems up for attack, and even launch the attack from within.

Part of the problem of protecting the Internet and computer systems is the imbalance between the difficulty of protecting versus the ease of destroying. The burdensome security procedures at airports are a good example. Someone tried to blow up a plane with a bomb in his shoe, and now US travelers take off their shoes when going through a security checkpoint.

Hacker organizations and individuals have shown the vulnerability of many systems. Many such attacks are a bit like stories of what Nobel Physics Laureate Richard Feynman reportedly did at Los Alamos National Laboratories when the atomic bomb was being developed: He left notes in top-secret safes (that he supposedly had no access to) saying "Feynman was here." These served to make security officials conscious of shortcomings in their systems. Hacker organizations whose motivation is showing off may act as early warning systems, although they might also inadvertently (or on purpose) cause significant harm.

Frank Jules, president of AT&T's global enterprise unit, speaking at a conference in November 2012, said that attacks on AT&T networks had doubled in the past four months and now tend to be more able to evade detection. He saw this trend as a business opportunity for AT&T. "Every chief information officer at major corporations that I meet wants to talk about security," he said. "I think this will be a $40 billion market one day." If major companies provide security solutions to other companies as a business, the synergy I've been discussing in the feedback from multiple customers will make the software better. The ability to buy security solutions rather than build them will make it easier for companies to address the issue. We are likely seeing a trend toward security being a growing commercial software opportunity, whether an embedded feature used to sell general software or as software targeted specifically at enhancing security.

One approach that can potentially deter a country from launching or sponsoring such attacks is to let it be known that we have tools to trace such attacks and a means of counterattack. In his October 2012 speech, Panetta warned that the US "won't succeed in preventing a cyberattack through improved defenses alone." He stated, "If we detect an imminent threat of attack that will cause significant physical destruction in the United States or kill American citizens, we need to have the option to take action against those who would attack us, to defend this nation when directed by the president. For these kinds of scenarios, the department has developed the capability to conduct effective operations to counter threats to our national interests in cyberspace."

A country might hesitate to launch a cyberattack if it believed such an attack would both be ineffective and traceable, and invite a response by economic sanctions or other means. Simply forming an agency to specifically defend against cyberattacks and launch counterattacks, as the US has done, has a deterrent effect.

But what about an attack launched by a terrorist group, where a counterattack by cyber technology is almost meaningless? Such groups might have difficulty launching an attack on infrastructure such as the electrical system. While they might hack into a power system, they would require considerable technical expertise to go the next step and sabotage equipment. If they did have such expertise, it is likely they would obtain it from a sponsoring nation. If that nation believed it could be traced back to them, then this becomes in effect an attack by the sponsoring nation and deterrence by fear of a counterattack is still relevant.

We can't simply assume the best. Developing defenses in critical industries is necessary to at least reduce the potential scope of damage. The *Wall Street Journal* reported in January 2013 that cyberattacks believed to be launched by Iran on financial institutions (presumably as retaliation for economic sanctions) were continuing. The article reported that the Business Roundtable, an industry organization, was attempting to get help from the government on countering such attacks. If a government sponsors the attacks, they are presumably sufficiently sophisticated that defense isn't easy, and skills to counter such attacks may be beyond those of the IT department of most companies. Government help might involve individuals at firms getting security clearances and getting advice from US government agencies.

9

Cultural Evolution

Rapid software evolution and the increasingly intimate connection between humans and computers changes our culture. The evolution of our culture—cultural evolution—proceeds increasingly quickly. It is evident today that cultural evolution is the main way human society is changing, with the changes largely external to our bodies. When I say that computer intelligence can almost become "part of us," I'm saying that the trends discussed in this book will become part of the culture that drives our society, which is as much part of what it means to be human as our bodies.

Cultural evolution is not limited to humans. Many examples of culturally established, persistent behavior in animals have been reported. Scientists have reported at least twenty cultural traditions in nine African populations of the common chimpanzee. Chimpanzee communities can adopt their own local customs and maintain their own multiple-tradition cultures.

Cultural evolution is a fundamentally important part of evolution. As I've emphasized, human culture is arguably evolving not just steadily, but at an accelerating pace through technology.

Human evolution has always been affected by our tools. Learning to build and control fires and devise weapons against predators are examples in human development that we all are familiar with. In the modern world, a multitude of new tools are created daily, and some become fundamental to the way humans operate. Automobiles and other modes of modern transportation have become essential to let us get places that would not be possible to visit regularly if we walked.

More specific to the software emphasis of this book, almost all of us have used a web search engine to expand the knowledge quickly available to us, far beyond the ability of any single brain to hold knowledge. And with mobile phones and other portable devices, this expansion of our human memory and analysis capabilities by orders of magnitude is available to use wherever we are. Tools enabled by software are a key aspect of our social interaction as well and an increasing component of cultural evolution.

Humans have evolved a complex society, a culture that involves people, practices, and the environment people create. The complexity is such that parts of our culture often surprise us. A notable example is periodic economic crises often caused by "bubbles," apparently irrational only in retrospect. An interesting question is whether technology itself (and software in particular) can eventually help us avoid the negative surprises, or at least moderate them. Forecasting of tropical storms is an example where technology such as satellites and weather prediction software have helped us prepare for what otherwise would be a surprise.

Cultural changes build upon themselves. Sometimes it seems as if the changes are random and uncontrolled, and, if so, we have little chance of making choices that change that cultural momentum. Before Darwin provided a theory of natural selection, it seemed as if the creation of species was also random. With the natural-selection model, we are able to see the patterns in the evolution of species and to understand how these apparently random events can be explained by a unifying theory.

If there were a pattern to cultural evolution, we could perhaps similarly understand it better, and, hopefully, learn to have it best serve the needs of society as a whole. One key difference between genetic and cultural evolution is that, while we can't select which random genetic changes will occur for genetic evolution to "select," we can exert some control over the changes in technology. Our experiments in technological change are not random, but controlled—at least, it is our intention to control them.

Humans have genetically evolved to deal with short-term problems fairly well, an adaptation sometimes called "fight-or-flight" reflexes in reference to our ancestors' need to make quick decisions in sometimes life-threatening situations. Rebecca Costa, in *The Watchman's Rattle: A Radical New Theory of Collapse*, notes that we are less equipped to make good long-term decisions, and tend to put off any short-term sacrifices that might aid us in the long term. She sees this shortcoming as the root of the collapse of ancient civilizations, and feels it threatens modern civilization if we don't recognize the challenge. Costa notes that, historically, when problems became so large that civilizations couldn't find a way to deal with them, they tended to substitute beliefs for logic, assuming that making sacrifices to the gods, for example, would save them.

However, I believe that cultural evolution can and to some degree has overcome this tendency to prefer short-term advantages to long-term interests. One way that culture addresses this tradeoff is money. Money is not part of genetic evolution of course; it is certainly cultural. It evolved culturally initially to solve short-term problems. Barter of goods was inefficient, since I might not have something you want to trade for what you have that I want. And, even if we have something to trade, your loaf of bread might not be worth one sheep. Money solved that problem and is a cornerstone of modern civilization.

But the use of money has evolved to make it a means of motivating our trading short-term benefits for long-term benefits. A simple example is a retirement plan. It's hard when you are young to start putting money away for retirement. Yet, by making money that goes into some retirement plans tax-deductible and gains in those plans through investments not taxable, the government encourages putting away money for retirement. While for some people, retirement plans are a company benefit and they don't have to make a real decision, many self-employed people are motivated to put away money that they can't take out without penalty for decades. The government policy reflected in the tax laws implicitly recognizes that those savings will be spent and taxed at a time that most recipients aren't working, a partial remedy for revenue issues with an aging population. I can't get into a discussion of the pros and cons of such a policy here, but it is a good example of how money makes it possible to "store" buying power much longer than we can store food, for example. And our attitude toward money is a cultural adaptation that we learn from childhood.

Software plays a role in long-term thinking as well. Certainly, software is the way most of our money is handled today, as bits in databases, but the connection goes well beyond that. Social networks and blogs, for example, allow people who believe in a danger such as global warming to find others with similar beliefs and organize to move society in the a direction to reduce the perceived threat. They may team with other interests, such as those concerned about reducing air pollution, to achieve both goals with similar methods. Regulations in the US for reducing emissions from automobiles have been effective, since an automobile manufacturer can't sell cars unless it satisfies those regulations. In the fuel economy case, of course, there is also a financial advantage in gas savings for buyers of cars with higher mileage. All of these cultural factors interact to move society in a direction that most would agree has long-term advantages.

Whatever it is that causes societal consistency can be rather fickle in the details. Effective design and marketing can cause rapid adoption of a new product category such as smartphones, for example. The rapid evolution of this category of device is an example of the acceleration of cultural change that software makes possible.

A focus of this book is technology's role in cultural evolution. Software is changing the evolution of technology—how will that change affect our future? The tighter connection between people and software and the accelerating rate of software change has deep implications for our society, including our economy, a subject I turn to in the next part of this book.

PART II

ECONOMICS

I've examined at length one impact of software moving quickly: The direction of the human connection with computer intelligence. That trend has an economic impact in itself—It's driving the next major innovation wave. Breakthrough technologies like the Internet seem to create major economic expansions, and the tightening human-computer connection is no exception.

A second major theme of this book is the changes in the economy driven by accelerating technological change. One area of change is directly related to the user interface trends that have been discussed in detail, the growth of "technology communities." A broader impact is the role of automation in reducing jobs, a topic whose importance is matched by its complexity. I begin with the simpler topic.

10

Technology Communities

As always-available access to computer intelligence and web resources becomes an increasing factor in consumer products and services, it is significantly affecting the computer/software industry in particular. The obvious results are in areas such as smartphone and tablet wars, with many companies struggling to come up with a "mobile strategy" and fighting vicious patent battles to protect their turf. These obvious conflicts reflect a deeper trend, a change in the way technology *communities* are formed, a result I attribute to the tightening of the human-computer connection and a desire of individuals to have computer intelligence and resources *always available* and *easily useable*.

To take a historical perspective on this trend, I note that there have always been communities in technology, often centered on formal standards, de facto standards, or open-source software. For example, the original "IBM PC" defined a hardware standard for PCs (based on Intel microprocessors), which Microsoft eventually took over in effect with the Windows PC operating system. A family was built around PCs running the Windows OS, including software packages running on the OS and hardware running Windows, with options from many suppliers, some of which became large companies. Apple's Macintosh OS created less of a community, since Apple made its own hardware; there are software options other than from Apple, however—Microsoft's Office suite and Adobe's Creative Suite being prime examples. In any case, with PCs, there are only two significant communities, Windows and Mac. Many companies—software and hardware—have historically flourished in this environment.

Users have benefited from a selection of compatible hardware and software within each community, with investment and creativity finding a large market because of the consistent platforms of each community; the job of software developers is easier if their products need be developed for a limited number of environments. The historical model based on PCs, however, is dissolving as mobile devices grow in popularity and begin to drive customer loyalty.

Another historical community was driven by mobile phone service providers and mobile phone manufacturers. The wireless service providers control the wireless channel and decide which phones they will sell, and this model put them historically in charge of the community around their particular service. While the phones made the direct connection to the subscriber, the early phones didn't have many features to distinguish them. Feature phones began to change that, with manufacturers like Research In Motion (RIM) with the Blackberry developing enough of a following that most wireless carriers had to carry the brand. Smartphones completed that dynamic, and wireless providers had to vie for the most popular phones, putting the phone manufacturers in charge.

In a clear analogy to the control IBM lost over the PC to Microsoft by ceding the user connection to Microsoft (resulting in IBM not even selling PCs today), users as this is written are largely in one of two communities—Apple's iOS or Google's Android, with Microsoft and RIM trying hard to present alternatives. Just as one has a choice between a PC running Windows or Macintosh operating systems, there is a choice of two primary communities. Manufacturers of the hardware (PCs or smartphones) can innovate on features such as design and battery life, but the direct connection to the user is controlled by the operating systems and the software supporting them, such as the search engines and map applications. Personal assistants or voice search are another user connection, and Apple and Google are attempting to provide their versions within their respective communities (although Google treats its assistant functionality as natural-language voice search).

Wireless service providers and phone manufacturers are trying to find ways to regain some of that control. For example, a Samsung Galaxy phone uses speech recognition from Nuance Communications to provide its S-Voice personal assistant application on an Android phone, although the Android OS provides speech recognition from Google as part of the Android operating system. S-Voice allows

Samsung some direct connection with the user without turning that connection over to Google.

Apple was the catalyst for a change in this community model for the mobile phone. It largely created a new community with its iPhone and iOS operating system. It forms much of the "community" itself, since it sells both hardware and software, is offering a cloud storage and synchronization service, and has an App Store that handles selling and delivery of apps. While app providers create a large community around the iPhone, Apple can replace almost any successful app with its own by bundling it with the phone, as they tried to do by replacing Google's map app (in a well-publicized debacle). Apple's Siri, as previously noted, is, in essence, a search product that uses Google search only when it can't bypass it.

Apple products and software work best when the customer has all Apple-branded products. While the most obvious examples of Apple's policy are the iPhone, iPad, iCloud, and Mac computers, Apple is widely expected to provide products wherever the consumer is, including its version of Smart TVs, maintaining to a large extent consistency of experience across the product line.

The Android community is more open, literally, since Android is an open-source operating system that anyone can use. However, Google controls its evolution and, as noted, add-ons that connect the user to Google's services. With Google buying Motorola Mobility, a phone manufacturer, the Android community may eventually look more like the iOS community. With these two having a large established base, it may be hard for another community to form, although I believe Microsoft has a chance of creating one, partnering in part with Nokia, and has a foothold on the desktop with their productivity software and in the living room with Xbox. Another possibility is that Samsung partnering with Nuance Communications could create their own extension of the Android community.

Android and iOS have made their way to tablets. The community model is even stronger with more than one device per user likely to use the same OS, since users don't want to learn two operating systems.

Before we examine the impact of Apple's success with this model, let's look at it in the context of the accelerating human-computer connection that this book has emphasized. Apple's advertisements emphasizing its voice personal assistant Siri send a message—"I'm easy to use." While some analysts have focused on the speech recognition aspects of Siri and limitations they see in its performance, I see Apple's positioning as being more corporate than specific to Siri. They are saying in effect: "We *connect* with *you*." Siri is just one example where Apple tries in all its products to provide an experience that brings you closer to the product, even the way the product looks and feels. Siri, with its natural language understanding and humor in dealing with off-the-wall questions, epitomizes this objective. But Apple takes the position that it has to control almost every aspect of user experience, in effect creating a community it dominates. Even with hundreds of thousands of apps in its App Store, it must approve all apps before it will sell them, and takes a cut of revenues they generate. Providing the applications built into the phone, e.g., the calendar/appointment application and contact list, Apple can make use of these applications by Siri seamless.

Amazon is another company creating a community of a somewhat different sort; it is the first place many people go, particularly to buy books, but increasingly to at least price almost anything. Amazon has a network of online retailers it supports in addition to selling directly, and provides the mechanism for customers to find products and buy them, with obvious advantages for partners. Amazon supports a community of publishers by finding them more audiences for their books, but as digital books become more prevalent, also has become a publisher itself. Amazon makes it easy for authors to publish a digital book and gives them a market and fulfillment mechanism.

While Amazon's community seems somewhat separate from those of companies such as Apple and Google, the communities are rapidly overlapping. Amazon is already selling its own electronic book reader and tablet. And Amazon is not alone in selling electronic content, Apple's iTunes being an obvious example. Amazon's community is not device-dependent, however; being a web service, it is available on any device, and popular enough that a device maker couldn't get away with blocking it. Amazon also makes its Kindle book reader software available on multiple platforms. Amazon's community could coexist with the iOS and Android communities, but could also face increasing competition as personal assistant software expands to find and price products.

Facebook has created another community. It is a social communication portal independent of the channel, with enough users that all channels must provide a connection to the portal. The company was seeking ways to extend this connection with users as this book was written. But they also may find themselves fighting the mobile communities; Google already has launched a Facebook competitor, Google+.

Microsoft has an advantage in that it already has a strong place in operating systems on PCs. Windows 8 is attempting to build on that and give a united feel to its PC and mobile platforms. They have their own community around Microsoft Office, with few people willing to climb the learning curve if they use that suite often. Their success with Xbox gives them an entry in the living room. They have lots of cash. They have the Bing search engine and the technology behind it. Microsoft Research has a long history of research in speech and natural language technologies. These assets and their apparent understanding of today's trends, as exhibited by Ballmer's statements, make it hard to discount their efforts.

There may be other companies capable of leading communities in human-connected technology. Research in Motion may be willing to license its new Blackberry operating system to other companies. In China, Baidu with its search engine or other Chinese companies may form China-centric communities. But the point is that there are clear communities forming, with a specific company leading each community, communities that will increasingly compete among themselves for our loyalty. If history is a guide, only a few communities can survive in the long run, and this is creating a major shift in the computer and software industries.

Another option for smartphones, Ubuntu, is based on the open-source operating system, Linux, and has its basis as an open-source alternative for desktop PCs. Linux is widely used on servers, but has little market share on PCs, although Ubuntu is available for PCs. The company Canonical announced in late 2012 that it is developing a version of Ubuntu for smartphones. Since any phone using it wouldn't be available until at least mid-2013, and apps supporting it beyond basic functionality would be limited initially, it may face an uphill battle.

Why do customers gravitate to these communities? Largely because the communities use a common theme in their user interfaces, using new services or products in the community becomes an intuitive experience if one is already using another component of that community. Once one becomes familiar with the user interface of these services, one doesn't want to learn another unless necessary. We are willing to learn something new if it builds on what we know, less so if it is unfamiliar.

Once you learned to type on a QWERTY keyboard, would you be willing to relearn a different layout? This is a telling example, since the QWERTY layout was designed to *slow* typing so that the keys on mechanical typewriters wouldn't jam; it isn't ergonomically optimal. The Dvorak keyboard layout is an alternative invented in 1936 that supposedly allows faster typing and reduces Repetitive Stress Injuries. It is supported by most PC operating systems, but few are willing to learn to use it. Touch keyboards still use the familiar QWERTY layout, even if typing on them is a completely different experience than a larger conventional keyboard.

Natural language is important because it is a universally intuitive interface. It can support increasingly complex features, yet let individuals accomplish tasks without a steep learning curve. Faster software evolution means that companies with natural language expertise can deliver more and more capabilities at an increasing rate, building on experience with users and making a challenge to an established community difficult. The work Google has done in optimizing its search engine shows the advantages of a large lead in a technology area.

Historically, the difficulty with "natural language" technology is that it wasn't natural. A good example is the use of overly structured speech recognition in call centers (telephone customer service operations). Call centers were cautious initially in their use of speech recognition, since early versions of the technology required limited vocabularies for accuracy over the phone lines. Companies tended to mimic their touch-tone structure, providing a series of prompts with limited options. Today, this is evolving with more companies essentially using a first prompt that says, "How can I help you?," but this option is more costly and difficult to manage than the structured approach, so it isn't yet common at this writing.

The natural-language approach is likely to become a requirement as consumers realize from mobile phone assistants that flexibility is possible. Eventually many consumers will start an interaction with a company's customer service through an app on a mobile device, possibly a specialized personal assistant as I have suggested earlier. Today, we largely struggle with unnatural automated customer service systems.

Call center use of speech recognition isn't the only example of technology using language in ways that are different than direct human communication. Web search such as Google or Microsoft's Bing historically operated on a list of typed keywords such as "United Airlines flight status" or "pizza 91316," not "natural" language. Google has updated its service at this writing to handle natural language queries (text or speech) such as "Is United Airlines flight 318 on time?", not only understanding the request, but providing the answer immediately, rather than just a web site listing. This evolution will continue as natural language technology improves and companies learn how consumers request information.

Whatever common user interface and hardware features define the community, every community will have a dominant company driving it, unlike technology communities based on standards, such as basic phone service. If that dominant company decides to deliver something itself that another partner in the community is currently delivering, that partner may find themselves with difficulty maintaining their market share. The dominant player also has a strong influence on the expansion of the community with new products or services that are consistent with established components of the community. The expansion can be through partnerships, internal invention, or acquisition. Without cooperation by the dominant company, an outside company would have trouble introducing something fundamentally new to the community. Patents, without patent reform, pose a further barrier to creation of a new community and to competition with an established community.

This is not to say there will be no choice for consumers. The choice will largely be about which technology community to join. There will be energetic competition, but competition between a few communities. It will become almost impossible to start a new community in an established area without some breakthrough innovation. In the future, communities will dominate much of technology that connects with people. The long-term economic implications are significant, but hard to predict.

It's not just sales to consumers that are at issue. Microsoft's Ballmer noted in the previously referenced note to shareholders that the "consumerization" of Information Technology in companies means the communities will extend to enterprises as well. A company is likely to favor a single community to ease its IT management challenges with Bring Your Own Device, for example.

While it may seem counterintuitive, this trend may favor startups that have an innovative product or service. There will be competition to bring such a product into a community. We are already seeing many small companies being acquired by larger companies at good prices even before they have built sales or even proved they can make a profit, sometimes with patents as an asset motivating the deal.

Some larger computer and information technology companies that don't control a community will have a hard time dealing with the change. Typically, a community may have a component that competes with such companies. It will be difficult to compete, other than on price, with a product that is part of an established community.

One might argue that the Personal Assistant Model (PAM) would ease transition from one community to another if all the communities simply gave you what you asked for in natural language. However, PAMs will learn as you use them, both improving the speech recognition and natural language understanding, making performance drop if one switched communities, until the new PA went through its "learning curve." A PA will also be tied to specialized community information sources in the cloud and on the device, something that would constitute another change if one switched allegiances.

A trend that can reduce the impact of communities may be companies providing bridges. Google and Amazon, in particular, provide some solutions that give a consistent experience across platforms. Bridge companies may be another category of power players in this technology-community environment, but only a few companies will have the power and services to be strong bridges.

Is this trend good or bad? It expedites a tighter human-computer connection and always-there availability because of the consistency of the community experience across devices. It gives the community members the market size to invest in continual improvement and to learn from a legion of users. The community trend could, however, eventually lead to some stagnation in creativity if a community rests on its laurels and considers that its customers are locked in. I hope if that happens that some innovator will come along and surprise the incumbents with a service good enough to build a new community. But there will probably be vigorous competition between communities that keeps innovation alive.

In the long run, the tightening human-computer connection will create many opportunities in technology, information science, and the creative arts. But there will be disruptions that could require some painful adjustments for some technology companies affected by the evolution of dominant technology communities.

11

The Economic Base

As this is written, the US and the world economy are struggling, with recovery from a recession proceeding at a slow pace, and with difficulty in even maintaining that slow recovery in the US and Europe. Productivity and automation in the economy play a key role in the recovery and long-term health of the economy. With software contributing increasingly to productivity enhancements, the same acceleration that we've seen in user interface innovation will take place in productivity trends.

There are two faces of productivity:

(1) Productivity enhancements due to technology have historically been key to economic growth and to providing an increasing quality of life for individuals, both by producing more product and services at lower cost or higher quality for the same cost, and by increasing overall production, making it at least theoretically possible to increase the number of jobs and/or the income those jobs produce. The higher productivity of US workers in manufacturing may create a renaissance in manufacturing jobs as productivity adjusted wages in the US become more competitive with increasing wages in developing countries, China in particular. Productivity improvements could also in theory make lower income for individuals more acceptable if that lower income bought more goods because of lower costs through the more efficient production. Higher productivity can potentially reduce the rate of inflation as well by making products cheaper.

(2) Some aspects of technology development and innovation can eliminate jobs (through automation of a process previously done by people) or even eliminate entire companies or

industries by innovation (what Joseph Schumpeter called "creative destruction"). If the workforce can't adjust to these trends quickly enough, the result could be stubborn unemployment.

These conflicting effects have historically favored the first factor and led to overall improvement in the economic well-being of the population, while requiring some parts of the population to adapt (often painfully). But there is no guarantee that this tradeoff will continue to benefit us without action to maintain the positive effects.

Even if the economy were currently healthy, the acceleration of change caused by software may be causing technology to move too fast for humans to adapt. This would cause a long-term problem in any case, as the jobs available didn't match the skills of available workers. As software is able to do more jobs that previously required humans, more jobs can be automated. One result, even if we weren't in a recession, could be that many workers take jobs that pay less than their previous jobs, leading to a drop in median incomes. (This may have already been slowly occurring, even before the current recession.)

Given that we are in a recession, companies that have laid off workers due to the recession are automating many jobs rather than replacing workers. It's easier to avoid rehiring workers than to eliminate jobs. There is a prejudice against hiring workers due to the associated costs and management problems that companies perceive come with workers, as opposed to automation (in essence, hiring computers). Thus, the trend toward automation is likely accelerated in recovery from a recession. This may be part of the reason job growth is currently anemic.

It's apparently rational for each company to follow this policy of not adding jobs if they can avoid it by automation. But viewed from a macroeconomic point of view, fewer jobs and lower median salaries cause a decline in consumption in the overall economy. Ultimately, this translates into lower sales on average for all companies. A decline in revenue reduces jobs further and encourages even more automation to retain profitability, creating a vicious downward spiral. We may very well be experiencing this difficult problem today.

While this scenario must be addressed, I should first note that productivity enhancements and automation appear to be currently creating some jobs by repatriating them to the US. Productivity can make those jobs more competitive with workers in other countries. In their 2012 book, *The US Manufacturing Renaissance: How Shifting Global Economics Are Creating an American Comeback*, Harold L. Sirkin, Justin Rose, and Michael Zinser note that in 2010 the average American worker was 3.4 times more productive than the average Chinese worker. Based on this and other factors, they assert:

> "Assuming current trends continue, by 2015, when adjusted for productivity differences, the costs to produce many goods in China and land them in the United States will be about the same as the costs to produce in the United States."

The authors' overall conclusion is that US manufacturing is in the early stages of making a comeback.

Further, it may be that some aspects of software and technology development are making it acceptable to accept a lower income. For example, if one can work from home because of computer networks and software, avoiding the cost of commuting and the time lost in doing so, the *net* income of an individual after work-related expenses may be higher than when the individual worked at a company at a somewhat higher salary. Inflation measures suggest costs are going up, albeit slowly, but those measures are based on the cost of a fixed basket of goods. If there is something in that basket I no longer need or need less of, such as gasoline because I'm working from home, inflation doesn't adequately measure my cost of living. If I now buy digital books instead of paper books at a reduced cost or download movies at a fraction of the cost of going to the theater, I'm getting a similar benefit at a lower cost. Perhaps technology enables me to be more connected with family members who live in another town (without travel expenses). I am not aware of studies on this point of "effective income." Technology efficiencies therefore may be somewhat of a countervailing effect, although probably minor, to the drop in median incomes.

Nevertheless, the historical role of productivity enhancements creating a stronger economy is not a rule of nature by any logic I can understand. This is not just a concern for the future. In America, median family income was less in 2011 than it was in 1989. As of October 2012, real median household income was only slightly above its October 2011 level, and 6.2% below the pre-recession level in October 2007. Low inflation and over-borrowing hid the potential distress created by this trend for a while, but we are paying for that over-borrowing as this is written. Very slow recovery in jobs from the recession that started in 2008 and stubborn unemployment rates suggest this is a legitimate concern. A large drop in consumer spending near the end of 2012 suggested that the current malaise has some core causes that aren't going away quickly.

Certainly, history suggests that it's dangerous to predict that automation has gone too far. In 1964, a committee with two Nobel Prize winners on it sent a report to President Lyndon Johnson warning that computers were likely to soon create mass unemployment. Any warning that we are approaching such a situation might be dismissed on the grounds that history has shown otherwise. On the other hand, the accelerating pace of technology development that software allows, as well as the increasingly complex tasks it can handle, suggest that we shouldn't dismiss this as an issue on the basis of history. (Most of history is in the past.)

The stubborn unemployment following the bursting of the mortgage bubble may not be just an issue of cyclical recovery. It may be that companies are avoiding re-hiring by choosing to accelerate automation instead, as suggested. It's much easier to avoid adding jobs in a recovery than to lay off workers in good times. The recession may just be accelerating a trend that was already underway, with middle- and low-income workers getting less of a share of economic growth than they have historically. If an open-minded examination shows that technology's contributions to the economy are tipping in the wrong direction (that the "destruction" in "creative destruction" is overriding the "creative" factor), we can take actions to minimize or prevent significant economic disruptions. In fact, technology itself is part of the solution; workers may benefit from the tighter integration with computers discussed in previous chapters, providing them skills and productivity they wouldn't otherwise have. As noted, US workers already have a productivity advantage over most countries due to adoption of past technology aids in doing their jobs.

The *Wall Street Journal* reported in April 2012 that companies in the S&P 500 stock index outperformed the rest of the economy due to "deep cost cutting during the downturn and caution during the recovery." As an indication of the greater efficiency, in 2011 the 500 companies generated $420,000 in revenue per employee, versus $378,000 per employee in 2007 before the slump. Unfortunately, those numbers translate into fewer jobs. The unemployment rate in the US was 6.1% or lower from July 1987 through August 2008 (dropping as low as 3.8% in April 2000), and has been above 6.1% since October 2008, hitting 10% in October 2009 before recovering modestly to about 8% in 2012. Unemployment in the 17 countries of the Eurozone was 11.7% in November 2012, including more than a quarter of Spaniards and 24.4% among young people. Unemployment was 10.7% in the 27 countries of the European Union.

The issue of technology advancing faster than the job market can handle is drawing more attention lately. For example, in their insightful book, *Race Against the Machine: How the Digital Revolution is Accelerating Innovation, Driving Productivity, and Irreversibly Transforming Employment and the Economy*, MIT researchers Erik Brynjolfsson and Andrew McAfee argue that we are already feeling the effect of digital technology in the economy. They speak of the "digital revolution," the increasing power of digital hardware and how software is getting smarter and doing more of what people are good at doing.

The April 21-27, 2012, issue of *The Economist* magazine included a special report, "The Third Industrial Revolution," that sounded a similar theme specific to manufacturing. The magazine summarized: "The digitization of manufacturing will transform the way goods are made—and change the politics of jobs too." The article noted that the price of pure labor would become less of a factor in the cost of products as digitally controlled machines create more products in whole or in part. "Most [jobs] will not be on the factory floor, but in the offices nearby," "the manufacturing jobs of the future will require more skills," and "the lines between manufacturing and services are blurring." The report opined, "Everything in the factories of the future will be run by smarter software. Digitization in manufacturing will have a disruptive effect every bit as big as in other industries that have gone digital, such as office equipment, telecoms, photography, music, publishing, and films."

The slow recovery from the recession that began in 2008 and led to worldwide malaise over time, despite heavy government intervention, is a warning that the problem suggested is already impacting us. Is the slow recovery of jobs in part because technology can be a substitute for re-hiring laid-off workers? Where are the jobs that created by software progress going to be, and are they a match for skills of today's workers? The change might be happening too fast for workers to develop skills for the jobs that are available, as suggested earlier; slower changes in the past allowed easier adjustment of the workforce through education and training.

I am perhaps being ambitious to approach this subject at all. For any economic theory, there is a counter-argument (and many people willing to make that argument). My general view as a technologist when I began writing this book was to show the potential of software to impact our society in positive ways, providing us with tools to expand our intelligence, skills, and humanity, as addressed in previous chapters. But as I dug deeper, I realized that the downside had to be examined—that there are real economic issues that have to be addressed to provide a balanced assessment. There is a question whether steady economic progress on average is possible or has reached some point where (a) there is a mismatch between technological progress and the jobs that match the talents of the current workforce and/or (b) there are simply not enough jobs left for the people who want them. I must examine this issue.

I won't just point out problems without suggesting possible solutions. Whatever policies this book recommends, the goal of steady economic progress with no dips is probably unrealistic. There will always be cycles, but they should be moderate, with a sustained upward trend. Economists usually argue that moderate cycles have some overall economic benefits; they can allow for corrections, where companies whose weaknesses were masked by overall growth are encouraged by downturns to introduce efficiencies that provide long-term improvements in productiveness or quality. Moderate cycles shouldn't cause a crisis that forces governments to step in and save companies that are "too big to fail."

Before I dig deeper into the role of technology in economics, I will set the stage by discussing some fundamental ideas that later discussion will assume.

The Feedback Principle

Adam Smith in his classic, *The Wealth of Nations*, focuses on the efficacy of allowing individuals to guide the economy, of letting individual choices of what to buy based on need (or desire) and price drive the marketplace. He recommended against governments controlling trade, preferring to let the marketplace adjust price and demand. His argument for free markets is typically viewed as the core of "capitalism."

One aspect of Smith's argument is that feedback from individuals through their purchases will cause companies to adjust prices and produce products to meet demand, feedback by what he called the "invisible hand." The idea of feedback is a core concept this book will employ in arguments beyond Smith's usage and beyond the economy.

In addressing problems in technology, society, and economics, we can use two fundamentally different approaches. One is feedback. With feedback, we monitor the results (the outputs) of a system or process to adjust that system or process; we apply adjustments to inputs or controls of that system or process in order to move the result in a desired direction. The other approach avoids feedback and simply uses assumed characteristics of the system (a "model") to try to control the system. Of course, an alternative is to let the system do its thing, but we are examining differing strategies for control.

The idea of feedback is intuitive. We might speak of a friend giving us "feedback" on a plan, meaning that the friend has listened to our explanation of the plan and reacted with advice that we may use to modify that plan. Most successful systems, e.g., the human body, don't operate by carefully calibrating each subsystem (heart rate, hormone levels, etc.). They operate using feedback: If a subsystem is out of balance with the rest of the subsystems, it receives a signal that tells it in which direction to change its activity. If our blood cells are not getting enough oxygen, we sense the lower level of oxygen and breathe more deeply.

A simple thermostat exemplifies feedback in technology. Suppose the thermostat controls a heating system. The thermostat is set to a desired temperature of the dwelling. It measures the ambient temperature to assess the impact of the heating system. When that temperature is below the target temperature, it either turns the heating system on or lets it continue to operate if it is on. When the temperature reaches the target temperature, it shuts the heating system off. The "feedback" is from the temperature caused by the heating system's actions.

What would be the alternative if a heating system didn't use a thermostat for feedback? The alternative is bizarre. We could enter parameters such as the size of the area heated, the amount of heat per minute generated by the heater, the capacity of the fan blowing out the hot air, use a sensor outside the house to measure the outside temperature, and model the heat transmission characteristics of the house's walls and windows, perhaps including some model of how often the doors are opened. The heating system could have a built-in computer that would compute how long the heater should run to move from the current temperature to the desired temperature given these parameters based on a model, and run the heater long enough to achieve that goal without feedback, that is, without actually measuring if this goal was achieved. Of course, this would usually fail and require the householder to manually turn the heater on and off, in essence becoming a human thermostat. The point of this extended example is that feedback can considerably simplify systems while achieving a better and more cost-effective result than more complex systems without feedback.

The complexity of the modeling approach illustrated with the thermostat can be compared with the difficulty of central planning of economies rather than allowing free markets. History has given us experimental cases comparing the two approaches, and the verdict of history has clearly been that extremes of central planning don't work in the long term. The example of a heating system without feedback perhaps suggests why central planning is ineffective and inefficient.

One danger of carrying the thermostat example too far is that a thermostat is designed to maintain a constant temperature. Feedback, more generally, could be used to maintain constant *change*, e.g., to control an economic process to maintain growth at a sustainable level without overly extreme cycles. Feedback could be used to support an activity that required constant adaptive changes to meet an objective, e.g., a robot vehicle avoiding obstacles. In fact, many engineers have been taught "feedback control," mathematical methods of achieving such adaptive control by design. "Automatic gain control" in the radio in your car, for example, monitors the average output voltage from the amplifier driving your speaker and, when the strength of the radio signal received varies, adjusts the amount of amplification to maintain the volume at the nominal level you have set with the volume control; such technology is so much a part of our lives we are seldom aware of it.

A properly designed feedback process can reduce the number of assumptions required to design the system and reduce the impact of inaccurate assumptions. However, improper feedback can also cause instability if it enhances a bad response rather than moderating or preventing it. A typical example you may have encountered in an auditorium is when the microphone is too close to the loudspeaker; the loud squeal results from the sound from the microphone being repeated with a slight delay by the loudspeaker and fed back into the microphone, amplified and thus creating a louder sound from the loudspeaker, creating a louder signal into the microphone, and so on. The rapid development of the mortgage crisis is an example of a few failures based on overinvesting in subprime mortgages causing a few failures that led to other failures that led to the value of assets at many firms plummeting rapidly, a feedback process propagating failure. The feedback principle can explain some good results and some bad ones. A colloquial way of putting the difference is "virtuous cycles" versus "vicious cycles."

In discussing software development, we commented that software is seldom perfect when first released to users; users use it in many more ways and in more computing environments than any programming staff could test. The users submit problem reports, and it is this *feedback* that allows iterative "de-bugging" of the software.

The result of properly applied feedback will tend to make the system self-correcting in its behavior. If we create economic or social systems that use the feedback principle properly, we have less need to fully understand their behavior. An example of the feedback principle in economics is tax policy; if you allow a tax deduction for home mortgages, for example, you encourage home ownership without requiring it.

Returning to Adam Smith, the invisible hand that causes prices to move toward their appropriate level is an example of the feedback principle. The collective actions of individuals vying for products they want from multiple merchants set the fair price. If a price charged by one merchant for a particular product were too high, individuals would provide feedback by buying from a competitor. Antitrust laws are designed to allow this process to occur fairly by insuring there are alternatives and that a consortium that avoids competition cannot set prices. The supply of specific goods coming into the market is a system in itself and is affected by feedback on how much total product is bought and the price it can demand.

A "planned" or "centralized" economy" attempts to decide the needs of a population and of external markets and organize the means of production to meet those goals. There is an obvious likelihood of a mismatch of needs and production and of pricing of products and services that doesn't match market demand.

In 2012, out-going Chinese Premier Wen Jiabao, in comments on Chinese public radio, called the country's state-owned banks a "monopoly" that has to be broken to allow freer flow of capital to loan-hungry smaller businesses. Translated, his comments to a group of businessmen to whom he was speaking were: "Frankly, our banks make profits far too easily. Why? Because a small number of major banks occupy a monopoly position, meaning one can only go to them for loans and capital. That's why, right now, as we're dealing with the issue of getting private capital into the finance sector, essentially, that means we have to break up their monopoly." This frank discussion perhaps indicates the degree to which Adam Smith's principles have permeated the world economy.

The economist John Maynard Keynes pointed out, however, in his 1932 seminal work *The General Theory of Employment, Interest and Money,* that capitalism in its purest form could lead to deep cycles, having the 1929 depression as a clear example when he wrote his book. He pointed out what I would call a feedback process that led to recessions and depressions, where a declining economy led consumers to save money and spend less (in part because they had less money), reducing market demand and driving the economy further down, leading to less income and spending, in a vicious cycle that might lead to a drop in the economy driven more by emotions than reason. He led a group of economists who felt that more intervention by government and central banks could prevent the worst depressions. With median incomes declining currently, even with more jobs being added slowly, there is a real risk of the vicious cycle Keynes describes driving our current economy down.

Using feedback in some situations can be technically impossible or economically infeasible. Traffic lights are an example of planning based on theory rather than feedback in most cases. In most intersections, the rate at which they change and which direction of traffic flow is favored are based on historical data. On any particular day, traffic accidents, street repairs, a big sale at a particular store, or other events that change the overall flow can make historical data invalid and result in traffic snarls, an effect we've all seen. Where it is economically feasible to measure traffic flow by sensors, feedback control has, however, been effective. In Los Angeles and some other cities, the rate of traffic entering from on-ramps on some freeways is controlled by stoplights whose frequency depends on freeway traffic flow rate at the location of the on-ramp. I was involved in early research on determining the optimal rate of on-ramp flow; we used a model that treated the traffic as fluid flow and the objective was to use feedback from measured traffic flow to avoid turbulence in the "fluid" at the on-ramp, an example of how engineering approximation can be applied to complex real-world situations and can be part of feedback control.

As noted earlier, feedback does not necessarily produce change in a good direction. We saw the bad effects of the wrong kind of feedback in the financial problems caused by bad mortgages. Ill-advised mortgages were given based on the assumption that housing prices would always go up. Mortgage officers received feedback in the form of commissions to close a mortgage, and banks received immediate profits by reselling these mortgages, in effect, as packages, presumably off-loading the risks. Hedge fund managers buying mortgage securities were motivated by bonuses for short-term performance; bonuses so large in some cases that even an eventual crash, if far enough in the future, would leave them with enough money to retire. Feedback was not related to the long-term consequences. If proper regulatory feedback forced banks to maintain proper reserves against the failure of these mortgages or penalized rating agencies for over-optimistic ratings, the result might have been different. This example deserves more detailed discussion than this book can attempt, but my point here is that feedback is a powerful effect for good and bad. Like any tool, safety requires it be used properly.

Feedback is powerful in part because, used properly, it can be "adaptive." That is, it can allow making small adjustments in one direction to see the effect and then making changes in another direction if the effect isn't what was desired. A series of small changes can in many cases steadily improve performance without the risk of a major gamble on a large change. With software, the adaptation is sometimes done in response to user actions, such as in Web search, where a user's choice between the options presented change future search results for others doing the same search. The wide use of web search provides millions of examples from which to adapt, without any one example overly affecting the result.

The Modularity Principle

Adam Smith used another idea that we will state as a general principle—modularity. He talked about the "division of labor." He was inspired by what became production lines in factories, using a pin factory as an example. In a production line, a series of workers each do a relatively simple repetitive task, resulting at the end of the production line in a complex product. A single worker would be hard pressed to learn all the skills required to build the final product alone. Even if it were possible for a single worker to acquire those skills, the production line would build products more efficiently for a number of reasons. For example, training new workers to fill a spot in the production line would be relatively quick and inexpensive relative to teaching them to build an entire product. Since each worker could become expert at a relatively small task, the quality of the final product would be better and more consistent.

Division of labor is the breaking of a complex task into simpler parts—what this book will generalize as "modules." I referred to modularity in an earlier discussion of software development. Technology uses modules to an extreme; a circuit board in an electronic device connects a number of components developed separately. The fact that a microprocessor can be used in many products means that it can economically be developed as a separate module, and every company developing an electronic product using a microprocessor doesn't need to develop a specialized chip for that product. It is evident that modularity is a key element of modern technology development, and part of the reason that most products are becoming digital. Because microprocessors can be used in many products, they can be sold in sufficient quantity to be cost-effective and to justify research leading to improvements.

Software is in layers, with an operating system and application programs being major modules in a digital system. The programs themselves make extensive use of subroutines to ease software development by breaking it down into simpler pieces. Some subroutines can also be re-used in multiple programs, making them much like electronic components.

Modularity is another basic principle that this book will appeal to a number of times. It is easier to create or understand a system or process if it is broken down into parts—modules—that we can analyze separately and combine to achieve the overall objective of the system or process. Modules can create efficiencies by being optimized separately and by re-use in multiple systems.

Modules can accelerate innovation when their use or creative combination makes possible a system or process otherwise not economically feasible. The use of a module in multiple products, services, or processes encourages its continuing improvement as a module, which in turn increases its uses and its utility and/or cost-effectiveness where it is already used—a virtuous feedback cycle.

A module can itself be composed of modules. And two or more modules can be visualized as "layers"—still modules, but thought of as a "stack" of modules with a lower layer interacting with a layer above it.

Modularity is another of those ideas, like feedback, that we understand intuitively and seldom think about. The human body is a set of modules. Our heart pumps blood. Our brain does many things, both consciously and unconsciously, to manage our body. Our kidney removes waste from our blood. Our digestive system takes nourishment from our food and disposes of the rest.

The Internet infrastructure is a module/layer that makes Web pages possible. The Internet is of course supported by a combination of modules that have other uses, e.g., servers and optical cables. Web site development tools are modules that make it possible to create and change web sites quickly.

In *The Pebble and the Avalanche: How Taking Things Apart Creates Revolutions*, Moshe Yudkowsky discusses how creating consistent modules (e.g., the Internet as a common way to connect any computer) can create revolutions, avalanches composed of mere pebbles. The principle is certainly powerful.

Like feedback, modularity can have its downside. Once you divide up a problem into parts, you may become stuck with a design that, with further analysis, may be the wrong way to break up the problem. If substantial effort has been put into each module, it may be difficult to change the design. One could argue that there are certainly signs of this problem in the organization of some governments and large companies. Some innovation has its roots in realizing that modules can be broken up and reconstituted differently.

Modularity is part of the explanation for the fast growth of technology in general as well as software in particular. W. Brian Arthur in *The Nature of Technology* notes that new technology is usually created by composing existing technology modules, although occasionally new modules (e.g., laser technology) are created. He traces this recognition that technology evolves in part by combinations of other technologies to Joseph Schumpeter, who argued that it was at the core of disruptive economic change. Arthur expands on this and related ideas to create a theory of how we should view technology and its evolution. As noted earlier, Arthur considers the "programming" of interactions between modules the way that technology develops rapidly. I can't do his deep analysis justice in a brief paragraph, but his point of view is consistent with the role of software as an acceleration of technology development.

THE ECONOMIC BASE

The principle of modularity is another reason that we will see advances in digital systems and software move at an accelerating pace. The more modules, the more opportunities to assemble them into better products or services.

Components of a healthy economy

Before delving into specifics of technology and the potential problem of automation going too far, I'll briefly outline basic assumptions I'm making about the economy that underlie any solution.

The rule of law

The law and its enforcement—the rule of law—is a requirement for a fully functional economy. The rule of law is another feedback process that makes a government work for its people, presuming the laws and their enforcement are fair. We want murderers and thieves to pay for their crimes; first, so that, once convicted, they can't continue to commit those crimes, but also to provide feedback to prospective criminals of what happens if they indulge in criminal activities.

This principle applies to businesses and their executives as well. An inequitable or corrupt legal system may be in some ways better than complete lawlessness, but in the long run a population and businesses need to feel that they will receive fair and equitable treatment if an economy is to flourish. Customers need to feel that they have legal recourse if cheated, and that this availability of recourse will motivate companies to treat them fairly and honestly.

Applying the law inequitably can cause discontent. In a poll in China commissioned by the Pew Research Center and conducted in mid-2012, half of those polled said official malfeasance is a "very big problem," up from the 39% who said so in a 2008 poll; another 35% termed it a "moderately big problem" in 2012. The Chinese leadership apparently is concerned about public opinion, and appears to be working to reduce corruption.

One reason that the US economy seems to be trusted by investors in other countries (as evidenced by their being willing to lend the country money at very low interest rates) is that we have a well-developed and respected justice system, supported by a free press that makes it hard to hide issues. There are certainly violations of the law by companies and executives, but wealth and influence seldom protects those violators permanently. Law and order is a fundamental requirement for a strong economy in the long run.

In what follows, the discussion assumes a functional legal and political system. When conservatives argue against government being overly involved in business, they don't mean that anything goes, including fraud and dangerous products. Technological innovations require an environment where the results of investment in those innovations can't be frivolously taken away because there is no legal protection.

Economics and politics

Historically, disruptive technology has had long-term positive effects, but has often been at odds with political objectives. Daron Acemoglu and James Robinson in *Why Nations Fail: The Origins of Power, Prosperity, and Poverty* (2012) argue in depth that the governmental structure and politics of nations determines economic success in the long run. "Extractive" political systems can cause technological disruptions that improve a backward economy temporarily, essentially by force, as when the Soviet Union initially moved from an agricultural to an industrial economy by state planning. "Inclusive" government structures allow the evolution of technology and the economy by providing opportunity for entrepreneurs to create new industries and efficiencies. Inclusive political systems allow individuals to move from one economic class to a higher economic class by skill, hard work, or creativity.

Disruptive changes can hurt a current elite class that has a vested interest in maintaining the status quo, but those changes are required for long-term prosperity for most of the population. When the elite has the power to prevent those changes, they will protect their positions, and economies eventually suffer. Acemoglu and Robinson use history as their laboratory, analyzing historical developments in country after country in depth to justify their conclusion that political systems are the main determinant of economic success, far beyond any intrinsic attribute of the country such as climate or resources. Robert B. Reich, in his 2012 book, *Beyond Outrage: What has gone wrong with our economy and our democracy, and how to fix them*, argues that we may be encountering a problem of an extractive system in the US, due to the growing influence of corporations on the political process through spending on elections, lobbyists, and hiring of politicians and bureaucrats after their term in government.

The authors cite what I would call a feedback principle driving the connection of political systems and their economies. They talk of how an inclusive political system can cause a virtuous upward spiral, while bad regimes decline in a vicious downward spiral.

There are of course distortions of this basic principle. Countries with significant natural resources such as oil can get sufficient economic leverage from those assets to avoid a fully open economy. This is a short-term condition that will fade as the resources decline and/or alternatives to those resources are adopted by a world that doesn't want to cede economic power to those countries. Acemoglu and Robinson claim that pressures on such economies from their own populations as economic discrepancies grow will also move those countries toward a choice between decline and inclusive politics.

The arguments of Acemoglu and Robinson may not be complete explanations of why nations fail, but they highlight the inseparability of politics and economics. Faster changes in the economy caused by technology acceleration may require the political system to adapt. An inability of a political decision to recognize the issue and to act in a timely manner will undermine the health of the economy and the political system itself.

I will propose some controversial remedies to the problems I've outlined. I hope that politicians and the corporate "elite" see the need to maintain an inclusive economy.

Government and non-government regulation

Government actions that affect the economy, as we have seen in recent times, are some of the most controversial. "Regulation" is unfortunately a politically charged word. But it perhaps shouldn't be. Law and order is a form of regulation by government; there is no law and order without government providing law enforcement and a judicial system. The real issue is the appropriate form and degree of regulation and the efficiency of its execution.

There are non-government regulations as well, usually set by industry organizations. Some "regulations" take the form of standards—regulations in the sense that they require certain characteristics if a product is to work according to the standard. These are certainly critical to today's technologies.

Government regulations

As noted, John Maynard Keynes advocated government involvement in the economy, in a position sometimes contrasted to the classical capitalist position that the free market will always lead to the best economic state. He argued for the use of fiscal and monetary measures to mitigate the adverse effects of economic recessions and depressions. This position seems to have been widely adopted by today's central banks.

Capitalism doesn't and never has implied that governments should not regulate commerce at all. The ideal of capitalism is based on the feedback principle—allow the actions of individuals making choices as to what they want and the price they will pay for it to determine prices and promote matching supply and demand. If there are factors interfering with the proper feedback, the right sort of regulations can eliminate those distortions. Monopolies certainly historically have interfered with that feedback by arbitrarily setting prices; monopolies that have the power to set prices irrespective of the marketplace interfere with the "invisible hand."

It's not controversial that we benefit from regulations that forbid child labor, false advertising, or the sale of poisonous food. Investors depend on regulations requiring public companies to issue audited quarterly reports. "Let the buyer beware" is not a viable policy; it would allow, for example, Ponzi schemes to go unpunished.

Lately, political debates tend to polarize between extremes, making consensus difficult on how much government involvement is effective and the appropriate forms of involvement. The arguments sometimes seem to become more slogans than thought, bringing to mind Einstein's caution about not making things more simple than necessary. Governments must balance over-involvement in the economy with the need to regulate certain behaviors.

The feedback principle would suggest that the best regulations are those that observe a measurable effect and expedite corrections based on those observations. An approach that depends on some assumed model of the economy are most likely to be inefficient and create unexpected consequences. Reducing costs of healthcare in such a way that it discourages individuals from occupations in healthcare (becoming doctors or nurses, for example), for example, would be a policy that would almost certainly create a future healthcare-quality crisis, particularly as the population ages.

Regulations that specify accounting standards for public companies reporting their financial results are an example of effective feedback. The government doesn't certify or create those reported numbers; they prosecute violations when they are brought to light. The regulations don't prevent bad behavior any more than laws against murder can prevent some murders occurring, but the laws discourage and punish bad behavior. Perhaps this feedback process is the best we can realistically do in most cases.

Taxes can be a form of feedback, as mentioned earlier. Taxes on short-term gains in the stock market higher than those on long-term gains discourage the former and encourage the latter. One can argue whether discouraging short-term trading should be a government objective, but it is efficient in promoting that objective. No agency needs to monitor your activity and you are not forbidden to execute short-term trades, but your behavior is affected by the feedback that you will be charged less if you hold a position long enough. Deferred taxes on income in retirement accounts encourages putting money into those accounts, arguably with an overall benefit to society as well as yourself, in that government is most likely to inherit the burden in some direct or indirect way if you have no money at retirement.

What's the role of software in this regulatory process? As an example, when the government requires financial reporting to a standard, it is creating information (a form of software by our broad definition) that can be compared across companies. Other software provides easy access to this information for individual and institutional investors, so that Adam Smith's invisible hand can allow reaction to that information.

Industry standards

Standards are a form of regulation when they are adopted by an industry, and contribute to more efficient use or dissemination of information. The World Health Organization (WHO) specifies the International Statistical Classification of Diseases and Related Health Problems (ICD), a medical classification list for the coding of diseases, signs and symptoms, abnormal findings, complaints, social circumstances, and external causes of injury or diseases. A physician or nurse specifies the correct codes for a particular patient, and the codes are widely used for such purposes as billing the patient based on treatments performed. This is a form of regulation by an industry rather than government, although a government can encourage use of such standards. For example, the US will subsidize the use of Electronic Medical Records, but requires "meaningful use" of such software, not just its installation.

The ICD standard is a good example of how software may help make such standards more effective. The current version most used is ICD-9, published in 1979 and based upon the state of medicine at that time. It has a relatively small number of codes, reducing its effectiveness in judging healthcare outcomes as well as providing less accurate detail for billing purposes. The main virtue of the limited codes is making their entry into a medical record easier.

The latest recommendation of WHO is the tenth revision, ICD-10, published in 1999. The US ICD-10 "Clinical Modification" of the basic standard has approximately 155,000 codes, relative to 17,000 codes in ICD-9, reflecting the diversity of treatments and diagnoses in medicine today. Healthcare organizations are just moving to this new standard, but it creates significant difficulties in accurately coding information for a patient, particularly since doctors often dictate reports in unstructured text that must be converted to codes, almost always done by a specialist other than the doctor. With ICD-10 having more than nine times the previous number of codes, it doesn't take much imagination to predict that this conversion will be difficult for many healthcare organizations and has the potential to create inaccurate or incomplete medical records. Software companies are trying to come to the rescue with solutions that can interpret unstructured and structured data and suggest correct ICD-10 codes. These entries will probably have to be reviewed by a human for legal reasons at a minimum, but review by a person can be much faster than requiring the person to originate the codes. The processing by software is also more likely to make the coding more *consistent* than a legion of humans, particularly considering the training required to perform this task accurately.

Moral hazard

Capitalism has been sometimes criticized as glamorizing greed. The opportunity to make money by creativity and hard work sometimes translates into more than taking advantage of the opportunity available; it can become taking unethical or illegal advantage of the system. That's why a strong system of law-and-order and sufficient regulation to discourage this "moral hazard" is critical.

Software can play a role in detecting fraud. A loss of hundreds of millions of dollars of customer funds at Peregrine Financial Group that came to light in 2012 in the US was apparently made possible in part by forged bank reports sent to financial regulators. The fraud apparently came to light when regulators pressured the firm to use a cloud-based software system for verifying accounts, Confirmation.com, that would most likely have uncovered the manipulation of reports. The firm's reluctance to do so motivated a deeper investigation.

Capital Confirmation, Inc., a private firm, provides the Confirmation.com service, which includes a process that validates the authenticity and authorization of each user, creating personal responsibility for any entries. The company advertises that the service is used by several hundred responding companies and over 10,000 accounting firms in more than one hundred countries; it provides an example of how software can help minimize moral hazard.

Charles Murray, in an editorial article in the *Wall Street Journal* (July 27, 2012), cited "the rise of collusive capitalism" as changing the view of capitalism as the economic model that "lifted the world out of poverty." One part of collusive capitalism, he said, is "crony capitalism, whereby the people on top take care of each other at shareholder expense." The more important component, he argues, is "corruption on a massive scale" encouraged by government regulations that can give competitive advantages if manipulated (e.g., "low-interest mortgages for people who are unlikely to pay them back").

There isn't a simple conclusion to be reached from this discussion. The main point I see when contemplating the problem of moral hazard is that tools that heighten transparency should reduce the likelihood of problems resulting from unfair exploitation of our economic system. Software tools can perhaps heighten this transparency.

The limits of monetary policy

Monetary policy is another aspect of economics that has recently been a key part of an attempt to revive the economy. Monetary policy has been used to make recessions less extreme. The US Federal Reserve has reduced short-term interest rates to near zero as this is written.

Manipulating the money supply by monetary policies has its limits. The approach is a bit like dealing with potholes in the road by providing subsidies to improve automobile suspensions. Monetary policies and suspensions can both smooth the bumps, but they don't actually remove the potholes. At some point, the potholes will get too deep, and the wheels will fall off. There are some intrinsic issues in today's economy, relating to the accelerating pace of technology advances, that require urgent attention beyond temporary patches.

What's good for the goose

In economics, what's good for the goose isn't necessarily good for the flock. An individual, for example, may be very prudent in cutting back on expenditures and increasing savings or reducing debt when the economy looks weak. But if most individuals in a weak economy do so, they spend less, revenues and profits for companies decrease, there are layoffs, and the economy gets weaker. As the economy gets weaker, individuals may get even more conservative, weakening the economy further. This is yet another example of the feedback principle, albeit operating in a vicious cycle. The only obvious way to get out of this cycle is to increase spending elsewhere to boost the economy, increase consumer confidence, and change individuals' reluctance to spend. Keynes would say that government spending is the only way to reverse the cycle.

Nobel prize winner Paul Krugman, in his 2012 *End This Depression Now!*, says it's just that simple. We should have learned from previous depressions what is necessary, and austerity isn't the solution. He cites "Keynes's central dictum": "The boom, not the slump, is the time for austerity." He cites one of Keynes best-known arguments for doing it now: "This long run is a misleading guide to current affairs. In the long run we are all dead."

A "Minsky Moment"

The economist Hyman Minsky described a key source of recessions and depressions, the overuse of debt. He noted that when times are good, it seems that loaning money is safe, but that this confidence builds on itself to a point of irrationality. When some trigger undermines that confidence, e.g., holders of subprime mortgages begin to default in large numbers, the high debt relative to resources has a compounding effect. Credit sources dry up, in part because of losses and in part because lenders get more cautious. Individuals and companies are unable to refinance the large debt in that environment, and failures and loan defaults multiply. This is another example of the feedback principle working in the wrong direction. The trigger that launches this vicious cycle has been called a "Minsky moment" by economist Paul McCulley. A simple and persuasive idea, it was largely ignored during the boom before the current recession. The message in the context of this book, perhaps, is that we must find solutions to economic growth that don't depend on over-extension of credit.

Internationalization

The internationalization of economies is another general topic that impacts the conclusions of this book. Today, companies and individuals compete to a larger degree with companies and individuals in other countries. While immigration and trade controls isolate a country's economy to some degree, the desire to get the economic benefits of cheaper labor in developing countries and to sell products to populations of those countries have reduced the trade barriers countries erect.

The "outsourcing" of jobs, particularly manufacturing jobs, to other countries by the US has caused pain for displaced workers. US workers cannot live on the wages that some workers in developing countries are willing to accept, or, for that matter, wouldn't accept some of the working conditions there. However, the trends in those countries are following past trends in developed countries, with increasing wages and improved working conditions. These trends, along with the cost of shipping products to another country and some quality issues, are—along with US productivity advantages— reversing some of the loss of manufacturing jobs in the US.

Taking a chauvinistic view of internationalization (for example, by creating trade restrictions) is shortsighted. All countries benefit from world stability and international economic growth. To take China as an example, its huge population is a both a challenge and a growing market. The Chinese government will face growing discontent in its population over time, as wide differences develop in incomes and resentment of corruption and an elite ruling class grows. If that resentment caused turmoil that the government handled with excessive force, it would create international political problems and potential disruptions in markets from which no country would benefit. Even worse, a historical response to internal problems is often to create an external enemy to unite the population, posing the danger of saber-rattling turning into an actual swordfight. I've chosen China as an example, since the country's economic importance is obvious, but the same arguments apply to other developing countries (and even some developed countries fighting economic problems).

While internationalization can create short-term problems for specific countries, including the US, these adjustments can also create a long-term healthy world economy that provides growing markets for the US, as well as fewer international political problems. Countries with steady economic growth that provide their populations with the possibilities for a better life for them and their children not only represent a more stable and rational world, but also reduce the likelihood of belligerent behavior. We all have an interest in that always difficult objective of "world peace," and economic growth can be a part of moving in that direction.

The cost of unemployment

As previously noted, unemployment in the US has remained stubbornly high during the recovery from the 2008 recession, hovering around 8%, up from 4.7% in December 2007, and not much below the 10% peak reported in November 2009. A measure of unemployment that includes a broader definition of unemployment, designated U6 by the US Bureau of Labor Statistics, measures total unemployed, plus all persons marginally attached to the labor force, plus total employed part time (for economic reasons rather than by choice), plus all persons marginally attached to the labor force (those who currently are neither working nor looking for work but indicate that they want and are available for a job and have looked for work sometime in the past 12 months). U6 was 14.6% in October 2012 on a seasonally adjusted basis, down slightly from 16.0% in October 2011.

Paul Krugman in *End This Depression Now!* opined that the best way to think about this continued slump is to accept the fact that we're in a depression. He quoted John Maynard Keynes' description from the 1930s: "a chronic condition of subnormal activity for a considerable period without any marked tendency either towards recovery or towards complete collapse."

Krugman emphasizes the hidden cost to individuals of a long period of low employment. He notes that those out of work use up their savings and often lose assets (such as a house), putting them at a long-term disadvantage even if they find work. He points out that a long period of unemployment on a resume can hurt job prospects permanently, leading at least to underemployment. Krugman notes that, statistically, those graduating from college during a recession have a permanent disadvantage in salary relative to those graduating during better times, often settling for jobs that don't reflect their education. The misery that an economic downturn causes is not necessarily fully reflected by unemployment statistics.

Krugman emphasizes Keynesian approaches to recovery, using government spending to boost the economy. But I am concerned that there is an underlying factor of over-automation that must be directly addressed if the recovery is to be long-term. I elaborate on this point in the next section, and I suggest a solution in a later chapter that I believe would be both effective and politically feasible.

12

The Role of Technology in Economics

Having provided a limited general discussion of economics that I will rely on in following discussions, I turn to the core issue of the impact of technology on economics.

A key theme, as indicated earlier, is whether automation is moving too fast for the economy and individuals to adjust. This book suggests that the acceleration of automation, and the increasing ability of software to do jobs previously requiring humans, may result in a net loss of jobs. Companies will respond by trying to become even more efficient as the economy shrinks, shedding jobs. Since the economy requires customers, fewer jobs mean fewer customers, which means fewer sales. You get it—a vicious downward spiral—the power of the feedback principle, but not in the right direction. A Minsky Moment created not by over-borrowing, but by over-automation. An example of "what's good for the goose" (a company trying to reduce costs) not being good for the flock (the economy).

Creative destruction

Much of classical economics discusses cases of equilibrium, the way, for example, that prices will settle at their natural level in the case of "perfect competition." Making an assumption of a stable system certainly makes creating a mathematical model easier, but doesn't reflect reality.

Joseph Schumpeter, in his 1942 classic *Capitalism, Socialism, and Democracy*, pointed out that instability was not only a characteristic of capitalism, but one of its strengths. He pointed out that constant change allows innovation and the growth of new industries. Technological innovation often makes older technologies, even entire industries, obsolete—a process he called "creative destruction," as previously noted. Few would argue today that such dynamics are part of a successful economy; in the extreme case, as Acemoglu and Robinson argue, if established industries and methods are protected by an elite, nations can fail.

Creative destruction has evidenced itself repeatedly in the past, with cases such as digital printers making typewriters obsolete and automobiles replacing the horse and buggy. As Kenneth Thurber says in the title of his book, "Do NOT invent buggy whips."

Of the twelve companies that made up the first Dow Jones Industrial Average in 1896—the industrial giants of the time—only General Electric remains. The modern economy also gives us many examples of creative destruction—and the accelerating pace with which digital technology allows it to proceed. One obvious example is the mobile phone market, where the growth in use of mobile phones has led to a situation where everyone seems to have one (even half of children aged 8-12 in the US, according to one survey), and some use it as their only phone. The entry into the market by Apple with its iPhone, by a company that had previously not sold one phone, upended the market. In 2007, when the iPhone with its iOS operating system was launched, about 65% of the wireless phones sold used Nokia's Symbian operating system. Research in Motion's Blackberry phone had about 9% of the market. In 2012, Apple's share had risen to about 23% of the market, Nokia was at about 8%, and RIM at about 7%, and RIM and Nokia were struggling to regain market share. Apple became the most valuable company on the New York Stock Exchange. Google's Android operating system was launched in 2008 and had captured over half the smartphone market in 2012 (within four years). An entire new industry creating smartphone apps flourished, with some 45 *billion* smartphone apps downloaded in 2012.

The smartphone story perhaps appears to be an extreme case. It certainly took much longer for automobiles to make the horse-and-buggy obsolete. But perhaps it is a typical case today in that it shows the speed with which creative destruction can occur in the digital era. The "destruction" in Creative Destruction is perhaps a frightening word, and for those workers laid off at RIM and

Nokia, it feels destructive. The "creative" aspect is the redeeming feature; the growth of Apple and Android smartphones apparently satisfied a need and is making life better for purchasers voting with their pocketbooks. The attractiveness of smartphones may reflect a fundamental trend in human connection to computers, as I have argued. It has created jobs at Apple and companies selling Android phones, as well as created new jobs for app developers and other supporting industries. The impact on overall job growth is difficult to assess. There is certainly an issue of whether the skills required for the new jobs such as app developers are represented widely among the population.

Automation is a more microscopic form of creative destruction, destroying jobs rather than destroying companies or industries. To look at one example, industrial robots are a case of automating jobs, replacing many jobs that required minimal training and education. Industrial robots on production lines handle repetitive tasks, move heavier items than humans can, and can be reprogrammed as the task varies, making it possible to use them to replace some production line workers.

With the rise of manufacturing in developing countries using low-cost labor, these machines have had less impact because of their high cost and limited flexibility. This may change with a low-cost robot such as the one being produced by Rodney Brooks, a pioneer in the field, with a robot called Baxter developed by his company, Rethink Robotics. With a cost of about $20,000 and programming largely through "showing" the robot what to do by moving its arms, Baxter can be used cost-effectively in relatively small production runs by smaller companies than those that use conventional industrial robots. It and similar devices could return some manufacturing to developed countries that would otherwise face the hurdle of overly high labor costs. On the other hand, they are another way that improved software (the key to the adaptability of Baxter) is allowing machines to automate more jobs previously requiring humans.

Automation is not just about replacing jobs or cutting costs. Much automation results in an improved product or service, improved in both performance and reliability. Production lines using automation can produce products that have minimal variation, where hand assembly or machining is more likely to create variable results. Computers let companies keep accurate records and manage services better. Slow restoration of power by the Long Island Power Authority after tropical storm Sandy was attributed by the media in part to dependence of the utility on humans to take calls and record problems, rather than using automated systems that could handle the increased load.

Older automated systems are often replaced by new automated systems because the new ones perform better or faster. Replacing a system doing a task with another system doing it better doesn't necessarily affect jobs. When a product can be produced more cheaply, yet with the same or better features, society benefits. Automation is not a villain, but it is a mixed blessing.

Productivity

"Productivity" is shorthand for a necessary aspect of modern economies. If the overall production of an economy is shared across its population, then increasing production will provide at least the opportunity for individuals to have more income and live better. Increased productivity is a necessity for developing countries with expanding populations; without it, there will be hardships for the poorest residents, and possibly even unrest that impacts the quality of life of more affluent segments of the population. Technology advances have historically been the main driver of productivity, although there are certainly other factors—management techniques and workforce education, for example.

The term "productivity" has the intuitive meaning suggested by the previous paragraph, but there are difficulties in defining it formally to capture that intent. Economists sometimes define productivity as a ratio of production output to what is required to produce it (inputs).

One measure of productivity for companies could be the value of products shipped and services provided (revenues) divided by number of employees used in producing that result. That measure, however, is difficult to compare across companies. A janitorial service is unlikely to generate more revenue without hiring more janitors. On the other hand, a business with a large software component can have a high ratio of revenue to number of employees. This ratio can also be distorted if part of the company's activities are outsourced, e.g., by using contract manufacturing.

Some companies report operating margin, the ratio of operating income divided by net sales in their quarterly reports. This number is somewhat related to productivity, but also related to the effectiveness of their marketing and competitiveness of their products. A premium brand, for example, can demand a premium price.

But, however measured, the ideal concept of productivity is clear: Producing a product more efficiently allows increased product volume and/or quality with the same resources. If the result of the efficiency is retained in the company, it can produce more profits, allows increased wages for each employee, and/or can be distributed to owners and shareholders. If translated in part to lower prices or higher quality at the same price, it can increase sales and benefit consumers. The two results aren't entirely independent: Increased product volume due to lower prices, for example, can result in efficiencies of scale and improve profits further.

Productivity can be estimated at the national level. Measures such as Gross National Product (GNP) or Gross Domestic Product (GDP) can be used to estimate overall economic production of a country. GNP measures the output generated by a country's enterprises (whether physically located domestically or abroad), and GDP measures the total output produced within a country's borders (whether produced by that country's own local firms or by foreign firms). Whatever the measure used, a percentage increase in the measure that was greater than the percentage increase in the population of a country over a given time period would seem to be a rough indicator of whether that country's individuals, on average, are better off at the end of the time period than before.

As an example of productivity growth, Sirkin, Rose, and Zinser, in *The US Manufacturing Renaissance*, point out that the US currently produces 2.5 times as much manufacturing "value added" (a measure of the additional value created by the manufacturing process over the raw materials used) as it did 40 years ago, using 30% less labor. This results in part from productivity of workers being higher with the tools they are using; productivity-adjusted wages show the US being about 33% lower than Japan and 25% lower than Germany because of the amount of product each worker can produce per dollar of wage. Sirkin, Rose, and Zinser in *The US Manufacturing Renaissance* estimate, assuming current trends continue, that, by 2015, when adjusted for productivity differences, the costs to produce many goods in China and transport them to the US will be about the same as the costs to produce in the US.

Lawrence Mishel, in an Economic Policy Institute Briefing Paper published in April 2012, noted that, since 1948, the value of product produced for each hour worked, averaged over the total US economy (inclusive of the private sector, government, and nonprofit sector), has grown steadily, up over 254% through 2010. Total production rising at a rate faster than population growth provides the opportunity for improved income for individuals, and the US population grew only about 8% during this period, much slower than the value of production. (The current annual growth rate of the US population is about 0.9% per year, including immigration.)

Increasingly, powerful software is improving productivity in areas that, on the surface, might seem beyond the range of what software can do. To take an extended example, consider what radiologists do. They look at an image (e.g., an x-ray or MRI) and decide if there is any abnormality and what the abnormality represents. This would seem to be a good example of human evaluation using pattern recognition that is unlikely to be replaced by computers. However, this is the kind of pattern recognition that computers can potentially do well, at least in the first step of finding an abnormality. An increasing number of images are stored in databases along with the radiologist's conclusions. Pattern recognition technology can simply take many examples of a certain type of x-ray (e.g., mammograms) and analyze this large database to determine what distinguishes those that are labeled "normal" from the rest. The analysis is mathematical, using techniques such as those in my book, *Computer Oriented Approaches to Pattern Recognition,* and later books on pattern recognition and machine learning.

The software doesn't "understand" anything in the human sense, it is simply providing a consistent explanation of the difference between normal images and images with suspicious findings. For example, a radiologist might note a shape in a breast x-ray that can indicate a possible tumor. A computer might highlight that part of the image as being anomalous. Such processing can be used to at least alert a radiologist of possible problems. A computer would not make a final diagnosis or recommend the next step, but such a screening can improve healthcare by reducing missed problems and/or speeding up the radiologist's examination of the image. In the long run, if the ultimate result with the patient is documented (whether or not the abnormality was actually cancerous) and those results statistically analyzed, the software may be able to provide a probabilistic estimate of the likelihood of the noted irregularity being serious. This issue has practical ramifications as recent research suggests that apparent irregularities found by radiologists on mammograms many times lead to unnecessary treatment. If an assessment of the probability of the irregularity being significant were available to the patient and the doctor, a more informed decision would be possible.

What does this mean for the radiologist's job? He won't be replaced by the pattern recognition software, but he will be able to process more images with more confidence in a given amount of time. He will be able to charge less for each image and maintain his income by reviewing more in a given time—improved productivity. This will reduce one aspect of healthcare costs, and it might avoid a potential shortage of radiologists due to lower compensation per case.

Today many doctors page through written notes in a paper file when reviewing a patient's history. Automation that took an electronic medical record for a patient and summarized key history and likely issues might help primary care physicians serve more patients. As noted, the US government is subsidizing "meaningful use" of the Electronic Medical Record in part for this reason. "Meaningful use" is a set of standards defined by the Centers for Medicare & Medicaid Services Incentive Programs that govern the use of electronic health records and allow eligible providers and hospitals to earn incentive payments by meeting specific criteria. This is an example of software expanding productivity in a key sector of the economy, preserving the quality of care by making a doctor more efficient and allowing the effective treatment of more patients by each doctor. Ultimately, we may have more objective evidence of the effectiveness of particular treatments through analyzing this data.

Productivity is a key factor in US competitiveness. Improvements in productivity result in part from the tools that US workers have available to them, including software tools.

But Mishel calculates that, while the hourly compensation of a typical worker grew in tandem with productivity from 1948-1973, that hasn't been the case since. Productivity in the economy grew by 80.4% between 1973 and 2011, but the growth of real hourly compensation of the median worker grew by just 10.7% (and nearly all of that growth occurred in a short window in the late 1990s). Mishel and Bivens reported in 2011 that the top 1% of earners secured 59.9% of the gains from 1979-2007, while the top 0.1% secured 36% of the gains, even though their numbers were of course only 10% of the top 1%. Only 8.6% of income gains have gone to the bottom 90%.

I've mentioned that productivity gains can make lower income more acceptable by lowering the cost of goods, thus making a given income go further. But the net effect of current productivity improvements is apparently not to reduce the cost of living in the US, at least by standard inflation measures, since inflation is growing, albeit slowly. Presumably, the cost of inputs to production other than labor (e.g., raw materials) are rising and are causing that inflation.

Software as a "nonrival" good

Paul Romer's classic paper, "Endogenous Technological Change," published in October 1990, went beyond the classical economic view that distinguished "public" goods supplied by governments and "private" goods supplied by the marketplace. Romer made a different distinction:

(1) "Rival" goods (goods with a physical embodiment that limits sharing, e.g., a candy bar, clothing, or a house). Once a candy bar is eaten, it can't be eaten again; only one person at a time can wear an item of clothing.

(2) "Nonrival" goods such as knowledge that can be stored in a form such as computer bits that allows, in principle, unlimited sharing. You can send a copy of a digital photo to a friend and still have it. Non-rival goods, such as a digital version of a hit song, aren't necessarily free; Romer called them "partially excludable" in that an artificial price *unrelated to the cost of production* could be set by an "owner."

This subject has deep implications for economics. Much of classical economics has been defined by the consideration of rival goods. The balance of "supply and demand" in classical economics usually assumes a limited supply. "Scarcity" is sometimes even used in the definition of economics. For example, in the introduction to his book, *Basic Economics* (4th edition, 2010), Thomas Sowell says: "Without scarcity, there is no need to economize—and therefore no economics. A distinguished British economist named Lionel Robbins gave the classic definition of economics: 'Economics is the study of the use of scarce resources which have alternative uses.'"

The classical economists treated technology and similar factors that couldn't easily be put in their models as "exogenous," that is, a factor outside of economics that could impact, but was not part of, economics. In the title of his paper, Romer emphasizes it is "endogenous," something that must be dealt with in economic models.

Software is a non-rival good, a form of knowledge representation that can be replicated as much as one wants. The fact that it has an increasing role in the economy reduces the importance of scarcity in economic models. Knowledge, one of the most important aspects of our economy, can today be easily shared and distributed. David Warsh, in his 2007 book *Knowledge and the Wealth of Nations*, emphasized the impact eloquently, beginning with a familiar saying:

> "Give a man a fish, and you feed him for a day. Teach a man how to fish, and you feed him for a lifetime. To which it now must be added, invent a better method of fishing, or of farming fish, selling fish, changing fish (through genetic engineering), or preventing overfishing in the sea, and you feed a great many people, because these methods can be copied virtually without cost and spread around the world."

TV shows and movies are not as fundamental to existence as food, but are an important part of modern society and economics nevertheless. When one watches a TV show, it doesn't disappear like fish when eaten. It still exists for someone else to watch as a digital asset. When one goes to a movie, the theater can be full or nearly empty, and the cost of projecting the movie doesn't change. When you watch a movie at home, there isn't even an incremental cost of presenting it, since you are using an existing TV set or PC. The movie is only scarce in the sense that the company holding the copyright controls access to it.

Amazon allows members of its Amazon Prime group, who pay a fixed price per year to be members, to watch thousands of movies and TV shows at no cost. Amazon obviously paid something for the rights to do this, but it costs Amazon very little to allow a movie or TV show episode to be viewed one more time.

By considering assets such as movies or books as intellectual property with owners, one could stretch the definition of scarcity. The owner can impose an artificial scarcity unrelated to the cost of manufacturing the product, Romer's "partially excludable" property. However, in today's economy, scarcity is not an effective concept for analysis, except perhaps for commodities such as oil or gold. Technology can even suddenly change supplies of scarce commodities, something we are seeing today with natural gas due to new techniques of getting it out of rock formations.

In fact, even physical goods have an element of information beyond the core product; we make decisions not on price alone, but on abstract information including perceived quality and brand. Take bread, for example, a product symbolic of fundamental needs. (Thomas Jefferson said, "A wise and frugal Government . . . shall not take from the mouth of labor the bread it has earned.") In one grocery store in the Los Angeles area, I found more than 25 varieties of bread being sold by brand and bread type (e.g., white, whole wheat, sourdough . . .), ranging in price per loaf from about $2.00 to $4.00. The price of a designer dress doesn't have much basis in the cost of materials and production; it reflects the creativity in styling and the brand. Apple seems to be able to demand a higher profit margin than competitors on a product simply because it has an Apple logo on it (and perhaps an edge in innovation and design). The varying perceived value in rival goods further reduces the ability of scarcity to explain the modern economy. Incremental value beyond the cost of production of a rival good is knowledge-driven (e.g., in the bread example, by brand awareness and advertising, as well as knowledge of the difference in nutritional value and taste). Let's call this the "information premium" for rival goods, an aspect of the movement toward a knowledge/software economy.

One of the most obvious areas of impact of non-rival goods is in books, music, and video. The ease of digital publishing is opening the way for creative talent that previously faced high hurdles to establish an audience. Writers can create a digital book by simply using an application on a web site to upload a text file and make it available though Web-based outlets. Today, books can even be printed as orders are received, as opposed to long print runs that assume a large volume of sales. And, while it still requires substantial capital to make a full movie or TV series, anyone with talent can get at least some recognition by uploading a short original video to YouTube.

The *Digital Music Report 2012* reported that digital music revenues to record companies grew by 8% globally in 2011 to an estimated $5.2 billion. Digital channels in 2011 accounted for an estimated 32% of record company revenues globally, up from 29% in 2010. Some markets now see more than half of their music revenues coming from digital channels, including the US (52%), South Korea (53%), and China (71%). International Federation of the Phonographic Industry (IFPI) estimates that 3.6 billion downloads were purchased globally in 2011, an increase of 17% (combining singles and album downloads). Plácido Domingo, the internationally known singer and conductor and chairman of IFPI, said in the group's *Digital Music Report 2012*, "Thanks to the amazing technology of the Internet, the audience for recorded music is fast-expanding across the world. Artists who might not otherwise find a way to make their music available can take advantage of the new ways to distribute music the Internet offers."

With the positives comes a negative. Digital entertainment can be pirated, and often is. IFPI and Nielsen estimated that, globally, 28% of Internet users access unauthorized services on a monthly basis. Around half of these are using peer-to-peer (P2P) networks, individuals sharing music; the other half includes web sites with illegal copies. In October 2011, Börsenverein des Deutschen Buchhandels, the organization representing German publishers and booksellers, reported that 60% of e-book downloads in Germany are illegal.

There is some evidence that easy and less-expensive access to digital material increases consumption of such material, a factor that can mitigate the impact of pirating. A *Pew Internet and American Life* report by Lee Rainie, Kathryn Zickuhr, Kristen Purcell, Mary Madden, and Johanna Brenner studied the impact of e-readers and digital books. Over one-fifth of American adults reported at the end of 2011 that they have read an e-book in the past year. The survey found that readers of e-books are more likely than others to have bought the most recent book they read, rather than borrowed it, and they are more likely than others to say they prefer to purchase books in general, often starting their search online. The average reader of e-books has read 24 books on average in the past 12 months, compared with an average of 15 books by a non-e-book consumer. The impact of easier publishing and more consumption is that opportunities for new authors may be one area of jobs that software is creating.

Types of nonrival goods

Software is of course one form of *expressing* knowledge, but it is also increasingly the mechanism by which knowledge is *delivered* and *utilized*. It is important, however, to make a distinction between types of nonrival goods to emphasize that software takes knowledge to a new level of accessibility. Some goods such as printed books are essentially *passive*—the information in them is hard to search and share. Some versions of knowledge—e.g., a digital representation of a book—are *active*. They can be quickly searched; sections can be marked digitally while reading the material for later retrieval; they can be copied quickly and shared electronically (e.g., by pasting a passage in an email and sending it, posting it on a social web site, or commenting on it in a blog); and they can even be *interactive*, e.g., with a link that leads to a web site when touched on a tablet computer. A web site can be passive, but most are active, at least featuring links to other web sites.

In another aspect of active goods, software can allow finding information by means that make the information more accessible. A database program is an example of software that allows finding structured data more quickly. You can probably find the full name of an acquaintance, for example, by searching for that acquaintance's company and first name in your contact database, for example, when the last name escapes you. Web search is perhaps the most used means of knowledge acquisition. Just type in a search request and get access to all the public knowledge on the Web.

A blog is active knowledge in another way. Readers can agree, disagree, or elaborate on a point made by the blogger, and those points of view are available to other readers.

A voice-activated personal assistant on a mobile phone or text-based search engine that allows an inquiry in "natural language" is perhaps one of the best examples of the power of an active solution in finding knowledge. Knowledge representation technology that condenses large sets of data, such as a web site, into more accessible categories or "answers" to inquiries is an evolving trend that makes natural language assistants more effective, as discussed in an earlier chapter. A mobile connected device that is tightly coupled to our human language capabilities and that can distill knowledge into a relatively compact answer is a hugely powerful tool. This type of human user interface is an example of a nonrival good that is at the boundary of what we can do effectively today, and an immensely important subject, as we have discussed.

The increasing importance of non-rival goods and the information premium in rival goods changes economic fundamentals. As David Warsh put it in *Knowledge and the Wealth of Nations*:

> ". . . once the economics of knowledge was recognized as differing in crucial respects (nonrival, partially excludable goods!) from the traditional economics of people (human beings with all their know-how, skills, and strengths) and things (traditional forms of capital, from natural resources to stocks and bonds), the matter was settled. The field had changed. The familiar principle of scarcity had been augmented by the important principle of abundance."

The human-computer connection as a major innovation

In a column by Jason Zweig in the December 15, 2012, issue of the Wall Street Journal, he quotes research by Robert J. Gordon of the Department of Economics, Northwestern University, that suggests we are in a period of slow US growth because there aren't innovations on the horizon that could drive growth to the degree that earlier breakthroughs such as railroads, electricity, the telephone, computers, and the Internet did. Gordon was quoted as saying that current innovation is "incremental," and not likely to drive the economy at the level we have enjoyed historically. Similar themes have been sounded elsewhere, but I cite this as a recent example of a concern about the dearth of breakthrough innovations.

I believe that the human-computer connection can be a new base for rapid innovation and economic growth, driven by innovation in technologies such as speech recognition, natural language understanding, knowledge representation, and the personal assistant model, with these resources being almost continually available due to mobile devices. The next chapter on software and jobs discusses one of the ways that these technologies can create new opportunities.

Because software is a non-rival good, its impact can grow very quickly. Unlike an innovation like railroads or the Internet, it doesn't require huge capital and time to have an effect.

It may seem that this expectation of a new technology base is similar to prognostications of computer intelligence taking over the world. As I have emphasized, it is far from that. I believe in a human-computer partnership rather than a competition. I believe that humans have unique skills, particularly in understanding other humans, and that the strength of computer intelligence is different (e.g., remembering and accessing a huge amount of information). The bridge between humans and computers was limited in the past, and we are just seeing the beginning of its full construction. The wide embrace of that connection by consumers and businesses can be the new innovation that drives economic growth.

Technology and capital

Another aspect of technology's role in the economy was emphasized by Karl Marx. He noted the difference between the hand mill and steam mill was that the steam mill required capital, creating a new dynamic in the economy. Schumpeter expanded Marx's example of the impact of technology further by noting that technology was creating "the propeller which is responsible first of all for economic and, in consequence of this, for any other social change." Schumpeter didn't agree with Marx's view of fixed social classes and class warfare being the result: "Supernormal intelligence and energy account for individual success and in particular for the *founding* of industrial positions in nine cases out of ten," he wrote.

Capital is available to startups, and this accounts for much vitality in economies such as that of the US. For example, venture capitalists invested $29.4 billion in 3,813 deals in 2007. Sequoia Capital, one of the largest and most successful venture firms, said on its web site that firms it backed (including Apple) are now worth a remarkable 20% of the value of the NASDAQ stock exchange.

Software creating hardware

Chris Anderson in his 2012 book *Makers: The New Industrial Revolution* argues that a new industrial revolution will be driven by hardware in essence becoming more like software. There are simple examples today: Some book publishers print a single book when they get an order, rather than using massive press runs. The evolving technology that Anderson cites allows a software-based design to be manufactured directly from the design by general-purpose computer-driven machinery. One type of such device is a "3D printer." A 3D printer creates a physical object by processes such as stacking one thin layer at a time in a material such as a plastic that then hardens. In cases where smaller production runs are sufficient, an individual sitting at a PC could produce a hardware product without a factory with such equipment. Such equipment is more likely to be used to create prototypes in the near term.

The US government is looking at 3D printing as a way to make the US more competitive against low-wage emerging economies. The government is providing $30 million to establish an "additive-manufacturing" research institute.

The trend described can amplify the impact of software on the economy. This book emphasizes the nature of software to allow rapid change. With the ability to drive more rapid change in hardware and to make hardware development available to more people, the acceleration of technology development is thus further enhanced.

Hardware becomes more like software, including a lower cost of launching and updating a product. Such technologies make some hardware more of a non-rival, easily replicated product. This trend is at an early stage as this book is written, but is likely to have a major impact in the long term, e.g., by making customization of hardware another way to increase the knowledge component of the hardware and thus its value.

Creating valuable companies

Software certainly can create value. One example that is perhaps an extreme, although there are other similar cases, is Instagram. Instagram was two years old with 14 employees when Facebook acquired it for $1 billion in 2012. Instagram created software for mobile phones that made it easy to share photos taken with the phone. Instagram is an example of a company developing a consumer-oriented service or product; as such, its product is highly visible to the general public and easy to understand.

While consumer-oriented companies like Apple and Google often get the lion's share of attention in the media, there are many software companies that provide software to businesses to help them be more efficient and effective, sometimes referred to as providing "enterprise" software. Such companies are not as well known to the general public, since they market to companies, not consumers, but they are receiving increased attention from investors.

For example, Oracle has a market capitalization of over $150 billion. It sells software solutions for companies, including database software that can handle very large databases. IBM has a market capitalization of over $240 billion, largely selling software and services to other companies. Salesforce.com provides cloud-based software supporting sales operations, as the name suggests, and has a market capitalization of over $20 billion. Selling software and services that help companies be more efficient and effective is a good business—good for the economy and often good business for the companies providing the solutions.

This section could of course be much longer. These few examples indicate the range of business opportunities in software. The point is simply that the role of software in the economy is matched by the business opportunity it represents.

The financial sector as a software industry

The financial segment of the economy is often viewed as a service industry, but it is essentially a software segment—it trades in bits. When you buy a stock, all that changes is an entry in a computer. Today, you might never deal with a person to buy or sell a security. You can sell the stock easily because the entry in the computer can be changed quickly using software that matches buyer and seller.

The money in your bank account is also just an entry in the computer. If you trade in commodities, the commodity may exist somewhere, but few that make such investments know where; no truck pulls up to your door and unloads ingots when you invest in gold.

Even mortgages, which seem to involve physical homes, are largely ledger entries and electronic documents to the financial industry. When a bank forecloses on a home, it is likely the officer taking that action has never seen the home and might not even be in the same city. The ability of financial institutions to "package" mortgages and create new financial instruments leading up to the subprime mortgage crisis depended on this abstract quality of homes as a financial instrument. The data describing the home become the reality to the financial institution, not the home.

Insurance has a similar quality. What you pay for life, auto, or home insurance is dependent on grouping of many similar cases so that statistics can allow setting a price that will most likely produce a profit from the group being insured. Whether it's a home, car, or life being insured, to the insurance company it's a record in a computer file. You may have put a paper copy of the policy in your files, but you'd probably have to go to court to collect if the insurance company's computer lost the record of that policy.

This observation of the digital nature of financial instruments isn't simply an abstract point. The financial sector has the characteristics that we have said make software a driving force in the acceleration of economic trends, both for good and bad.

The good is fairly obvious. Without computers, you'd never be able to trust your bank balance or use a credit card. Stock prices would still be set by traders in a big room yelling offers to buy and sell to each other.

The bad is perhaps also obvious in hindsight. The financial crisis that began in 2007 and is continuing as this is written is generally considered to have been driven by the bursting of a housing bubble and the over-awarding of subprime loans. It more accurately was caused by the use of abstract instruments that allowed mortgage lenders to pass on the risk of a questionable loan or believe they were insuring against the risk. Lenders were motivated to make loans that they might not have made if they thought they were absorbing all the risk of those loans. The buyers of securities depending on those mortgages perhaps assumed that the originators had been more prudent in assessing the credit-worthiness of the borrowers or the value of the properties. Beyond that, the layers of securities based on other securities hid the risk from many investors. Both individual investors and companies invested some of their money in mutual funds and hedge funds. Mutual funds and hedge funds had more invested in risky mortgage securities than they should have. Insurers such as AIG added another layer, supporting the rationale for buying risky securities by implying they were insured, when the capital behind the insurance was insufficient to support a significant across-the-board failure.

It's as if a company specializing in earthquake insurance assumed that an earthquake would never happen (or perhaps that the government would bail them out if it did), and made no effort to set up capital reserves to pay if the earthquake happened. The layers of mortgage instruments compounded to a level where one failure would reverberate across the entire system. Since each layer added another way for financial firms to move bits around and take a fee for doing so, there was little motivation for firms to police themselves. Because finance is software, it was relatively easy to create the layers quickly, and for them to have more of the characteristics of software in that "bugs" weren't discovered until they caused a problem.

Collateral Debt Obligations (CDOs) were an example of software with hidden bugs, to continue the analogy. A set of sub-prime mortgages, for example, could be put in a package—"securitized."

As a single package, they would have to have a high risk rating, so a trick that changed junk to gold was invented. The bad loans were separated into "tranches" (slices) using three groups in the simplest case, and investors could invest in separate tranches. The most risky tranche ("equity") got the larger return by taking the direct risk of mortgages not being paid off; the second tranche ("mezzanine") got a lower return because in theory the equity tranche was paid last and took losses first, but the mezzanine tranche was next in line. The third tranche ("senior") was first to take profits and last to take losses. By this sleight-of-hand, the senior tranche often got an AAA rating, despite being dependent on BBB-rated mortgages. Perhaps the senior tranche was indeed safer in some sense, but the value of *all* tranches evaporated rapidly when the sub-prime crisis hit. You can't sell something when there are no buyers.

The ease of creating new financial "software" has also led to a host of "shadow banks," hedge funds and other institutions that act much like banks in that they invest other people's money. The distinction is that these organizations don't get their money from depositors, but from "investors" or other institutions, including banks and other shadow banks. They aren't protected from runs with deposit insurance as real banks are, and they aren't subject to the same regulations. The usual practice has been to use a lot of leverage, often borrowing twenty times the amount of the funds invested, thus getting almost twenty times the return that the invested funds alone would have received (if there is a return). Those borrowed funds were often invested in other shadow banks, which leveraged those investments. If this sounds overstated, I urge the reader to check out more detailed assessments, such as Roubini and Mihm's *Crisis Economics*. The shadow-bank house of cards is an example of moral hazard caused by a "hear no evil, see no evil, speak no evil" mentality.

The layers of financial instruments led to a domino effect. The liquidity that was eliminated by the disappearance of much value from the computer entries in financial firms led to further problems. Consumers suddenly had less money because the entries in computers that said what their mutual funds or stocks were worth suddenly declined at the same time the equity in their home and their ability to borrow against it evaporated. The recession might have become a depression if the federal government hadn't stepped in and added huge amounts of liquidity to the system (by changing a few computer entries at central banks that allowed the changing of a few more computer entries at the financial institutions that were bailed out).

The financial industry depends on real things being translated into bits, and it depends on software to manipulate those bits. Occasionally the software fails, as it did in the flash crash of May 2010, as discussed in the section on software reliability. But the point we are emphasizing in this section isn't the dependence of the sector on software (since all industrial segments depend on software today), but the fact that the financial sector *is* software. It can be changed rapidly, so it exhibits the general trend of technology changes accelerating. Can financial institutions keep up with this rapid change? Will there always be unintended consequences (bugs) as new financial instruments—new software applications—are invented? Will the automated trading used by some investment funds cause aberrations in prices of financial instruments that don't reflect their true value? Can regulators monitor these complex interactions?

If a mutual fund is only measured by annual performance, and a fund manager knows that beating its peers by a few percent in a given year will cause a major influx of funds, the manager is motivated to take risks to get that few percent. There would be even more temptation to take increasingly risky bets if a fund was falling behind the indexes.

Is there some way that investors could be more sure what they were getting? It would be interesting to see if the temptation to take risk was reduced by financial firms' monitoring and reporting risk. A firm that reported risk in a credible way would have a competitive advantage over firms that didn't. If investors were convinced that one fund was safer than another, money would probably pour into that fund. Investors actually absorbed negative interest to invest in German bonds at one point in 2012 to reduce the risk of losing their money, paying for the right to give the German government a loan. The markets recognize lower risk and reward it in the price of securities reflecting that lower risk, so it should be good business to measure and report risk accurately. I'm certainly not qualified to recommend a way to measure that risk, but, if a few firms found a credible method of doing so, it would be a competitive necessity for most other firms to do so.

While a given fund might find it hard to credibly estimate its own level of risk, perhaps independent firms that develop methodology do so for a fee will flourish. There are, of course, rating agencies, but their credibility was damaged by the last crisis. Part of the difficulty is that any outside assessment will largely depend on periodic reports from the entity being assessed. In theory, software could report almost daily some commonly accepted measures of risk, just as a mutual fund reports its value daily. It's likely that software that can adequately measure risk or confirm financial reports will play a large part in any safety-driven revolution in finance.

Another aspect of the rapid change possible because financial institutions are essentially software is the setting of values for financial instruments. Take stocks, for example. We know that a change in perception can drive a stock up or down quickly. Some of this may be real value, e.g., when Schumpeter's creative destruction makes a loser out of a winner or vice versa, with the competition in smartphones a dramatic example. In 2012, Research in Motions's stock price had dropped to 5% of what it was valued at its five-year high in 2008. One could argue that this is basic capitalism at work, Adam Smith at his best, with buyers and sellers negotiating the value of a company. But investors in general aren't assessing a company by checking out its factories or talking directly to officers to assess their intelligence and diligence. They are depending on assessments in the press, financial reports, and parameters such as price-to-earnings ratio, sources that can change more quickly than factories or management quality. Sometimes the "knowledge premium" in stock prices turns out to be a "lack-of-knowledge" discount.

Measuring success of financial firms by performance can be misleading. Simple statistics says that if a number of investment professionals make random decisions on investment, some will beat the average performance of the categories of financial instruments in which they are investing in a given year and some won't, whatever their investment methodology. The winners will get investors to put more money into their funds, and the losers can tell their investors that the strategy will pay off next year. As Paul Krugman put it:

"Hedge fund managers, for example, get paid both fees for the job of managing other people's money and a percentage of their profits. This gives them every incentive to make risky, highly leveraged investments: if things go well, they are richly rewarded, whereas if and when things go badly, they don't have to return their previous gains." (From *End This Depression Now!*).

In his 2012 book, *The Hedge Fund Mirage*, Simon Lack calculated that, over the past decade, investors in hedge funds would, on average, have done better putting their money in safe, low-interest Treasury bills, suggesting that the model of random winners and losers is probably fairly accurate.

Despite all the potential problems, the financial system is critical. Without cars, there wouldn't be automobile accidents—that fact isn't likely to cause us to abandon cars. How do we minimize auto accidents? Safety features on cars, and traffic laws with penalties. The solution for a better financial system is the same. We need measurements such as the reserve capital of banks as public information, with penalties for falsification. We need prison terms for those who break the law. Widely reported trials of previously prominent individuals who believed they were above the law (e.g., who were found guilty of insider trading) can reduce the likelihood of others making that assumption.

Software can help in providing information to regulators and investors that highlights problems before they do major damage. The US Securities and Exchange Commission (SEC) approved a plan in July 2012 to track all US stock trading with a computer system, partially in reaction to the 2010 flash crash, to get a better assessment of the rapid-fire computer trading that has become prevalent in US markets. The proposed system would track orders, cancellations, and executions of all US-listed stocks and options, across all markets, delivering the data to the SEC in a uniform format by the next trading day. The exchanges and the Financial Industry Regulatory Authority, Wall Street's self-policing body, would build the system.

The financial system, being one of the most automated industries, can shed jobs quickly when there is a downturn. For example, in

October 2012, New York state comptroller Thomas DiNapoli released a report noting that the securities industry in New York lost 28,100 jobs during the financial crisis and added back only 7,900 during the recovery. Nevertheless, the average salary, including bonuses, paid to security-industry employees in the city was $362,950 for 2011, up slightly from the previous year, according to the report. I'd like to see how classical economics would explain that discrepancy.

13

Software and Jobs

What does the evolution of software mean for the availability and nature of jobs? Brynjolfsson and McAfee, in *Race Against The Machine*, say that they believe that technological change is the major impact on the job market "because the pace has sped up so much that it's left a lot of people behind. Many workers, in short, are losing the race against the machine."

Software is allowing technology to evolve ever more quickly and to do things previously requiring humans, and humans intrinsically adapt more slowly. We need to examine if "creative destruction" can become "job destruction" to the point where the overall effect is a smaller economy as people without jobs buy less, encouraging more automation to reduce costs, reducing jobs and disposable income further. This vicious cycle—driven by the wrong kind of feedback—could drive the overall economy down.

I'm not saying that technology evolution and increased efficiency should be deliberately slowed (if that were even possible in competitive situations). But we need to understand this process and see what societal or economic changes are possible and necessary to avoid an economic crisis. Unfortunately, I believe we are already seeing evidence of the problem in the slow recovery from the current recession.

We'd all like economics to be a science with clear, provable methods, but it isn't. Engineers have the luxury of using laws of physics that don't change over time; they can usually agree, for example, whether a bridge will hold the load for which it was designed or that an electrical circuit will do what it was designed to do. While these "laws" sometimes turn out to be approximations—Einstein proved some of Newton's Laws don't hold up at speeds approaching the speed of light—they are accurate within a range of conditions that are unlikely to be violated in the systems being designed. Economists are dealing with social and technological environments that change over time, and, as this book has argued, can change rapidly. Almost every book on economics is forced to explain conflicting arguments and theories, and most books by economists argue against theories of other economists.

Part of the difficulty is that one can't do repeatable experiments in economics as one can in physics. The underlying societal system changes rapidly enough that what one observes in the past may not be relevant to the current conditions, and any experiment in macroeconomics requires a long time to evaluate. I make this point because I believe we are seeing a fundamental change in basic assumptions because of the rapid acceleration of technology development. We must try to understand this change and how to manage it.

Productivity and automation

Productivity is an economic concept defined rather loosely as a ratio of production "output" to the "inputs" required to produce it, where one of those inputs is labor, as previously discussed. Historically, increased productivity created by technology has been credited with improving overall economic health by generating more product with the same resources or lowering the cost of products. Improved productivity can allow total output of a country to expand faster than the population, making more wealth available per person. Increased disposable income can be used to buy more products. If these conditions hold, the economy expands in a virtuous cycle.

However, when productivity is called "automation," it sounds more ominous. Automating activities done by humans, even if more efficient, has the prospect of destroying jobs. Economists such as Schumpeter suggest that the effect of this "creative destruction" is a net positive, creating more jobs in the end. Some economists (e.g., Jeremy Rifkin in 1995 in *The End of Work* and Erik Brynjolfsson and Andrew McAfee in 2011 in *Race Against the Machine*) have warned that there are limits on this benefit, and we could reach the point where automation creates a net loss of jobs.

Statistics I previously cited show that the growth created by automation can also be distributed unevenly, with higher-income groups getting more of the benefits than lower- and middle-income groups. Since many of the most wealthy individuals get wealthy by earning money on the capital required to invest in automation, one could stretch a point and note the development is coming dangerously close to Karl Marx's expectation of fixed classes (an ownership class which controls production and a labor class which produces the goods), with the danger of some form of class warfare. Thomas Friedman, in an editorial discussing the 2012 Presidential election results in the *New York Times* (dated November 7, 2012) predicted that "the biggest domestic issue in the next four years will be how we respond to changes in technology, globalization and markets that have, in a very short space of time, made the decent-wage, middle-skilled job—the backbone of the middle class—increasingly obsolete. The only decent-wage jobs will be high-skilled ones."

The impact of automation may pull down income for an increasing share of people as the automation uses software to do more of the work that has typically required human intelligence and skills. This "smart" software can already do things like language translation or speaking with callers at a customer service center, once the sole purview of humans. Industrial robots are doing some things that previously required human dexterity.

Historically, changes impacting jobs were gradual. The transition from an agricultural to an industrial economy, for example, was slow enough that it allowed time for migration to cities and for education to prepare workers for another type of work. Faster change in available jobs could create an imbalance that creates a long-term problem not easily addressed by a short-term stimulus such as monetary loosening. Further, there simply might not be *enough* jobs, whatever their nature, to match the number of people seeking work.

The problem seems to be occurring *now*. Massachusetts Institute of Technology economists David Autor and Daron Acemoglu examined in detail the changes in the types of jobs the economy is supporting (reported in the *Wall Street Journal* in July 2012). The researchers segmented the US workforce into 318 occupations, ranked by skill and education. Between 1989 and 2007—before the recession—they found only a 5% increase in routinized production, machine-operator, and clerical jobs that could be automated in part—but a 36% increase in personal-service jobs that are difficult to automate (e.g., hairdressers, nursing aides, and food preparation), and a 40% increase in jobs requiring the most education and skills (e.g., managers, professionals, and financial professionals). An updated analysis by Autor and Acemoglu, according to the *Journal* article, showed that, between 2007 and 2010, while the total number of jobs in the US fell by nearly 6%, the previous trends continued. The number of middle-skill jobs in the US, those most impacted by automation or offshoring, fell by 12%, while the number of high-end, high-education jobs fell by only 1% and personal-service jobs exhibited 2% growth.

Statistics on company investment in equipment and software provides further evidence that automation is replacing some jobs. Investment dropped, as one might expect, in the early stages of the 2008 recession, as business revenues rapidly contracted. As the economy slowly recovered, investment in equipment and software increased in every quarter from the last quarter of 2009 to the second quarter of 2012. (Investment appeared to slow in the last quarters of 2012.) During this period, jobs grew very slowly. Since growth in Gross Domestic Product during this period was dismal, it is unlikely the capital investment was necessary to support increased production; it was most likely investment in automation to replace workers rather than rehire them.

In the previously cited 2011 *Race Against the Machine*, Brynjolfsson and McAfee sound a similar theme. They note that while many measures of economic activity recovered to near-pre-2008-recession levels, employment did not. In particular, they note that investment in equipment and software returned to 95% of its historical peak, a fast recovery relative to previous recessions.

Jobs that are physically challenging or dangerous are particularly subject to automation, even at a high cost. For example, machinery that automates mining almost completely is being adopted. The need is partially motivated by scarcity of resources, forcing deeper (and thus more dangerous) mines. For example, the owner of a nickel mine is Canada has spent nearly $50 million to install and test equipment from Rail-Veyor Technologies Global. The system is run by an operator above ground using computer controls. It digs out and loads nickel ore into containers that run on a rail. When full, the conveyors take the ore through the mine and above ground. No miners are involved.

The rise of online shopping is another area where automation may reduce jobs. Web shopping is software-driven and can do things that a salesperson at a store might not be able to do. For example, online vendors use advanced software techniques to suggest what "shoppers like you" might want or what other things were typically bought in conjunction with the product you are buying. Whether you consider this an annoyance or a service is perhaps an individual preference, but it appears to increase sales. The process of online shopping is almost entirely automated, from showing you alternatives to sending the order to a fulfillment center. While this might increase jobs in distribution centers and shipping operations, it apparently reduces jobs in retail locations, judging from trends reported by companies that generate most of their revenues from such stores. It is likely that the net result is a loss of jobs as sales volume in local stores shrinks.

Some jobs are of course being *created* by technology. There is a continuing need for expertise in computer science and programming. Web site designers and developers are positions that couldn't exist before the Web. Technical support agents reached by phone or chat or in computer stores are needed because new technology requires support. Writing mobile apps is a new class of job. Occupations driven by new technology are a source of some new jobs, but typically specialized ones that require a particular aptitude as well as training.

An earlier assessment of the distribution of jobs in the 2004 book *The New Division of Labor: How Computers are Creating the Next Job Market*, by Frank Levy of MIT and Richard J. Murnane of Harvard showed a similar trend even before the recession, the "hollowing out" of the critical center of the economy. Levy and Murnane postulate that jobs where people essentially follow a set of relatively well-defined rules that can be implemented in software, such as mortgage loan agents that grant home loans based on numerical criteria about the home and homeowner, will decline in number and in the wages they can demand. They surmise that jobs that require "pattern recognition," such as a doctor diagnosing an illness or a service technician repairing a malfunctioning automobile, are the type of jobs that will grow in number and value. Such pattern-recognition-based jobs can be relatively mundane; for example, a maid, food server, or gardener is unlikely to be automated due to the complexity of what they do (as well as the relatively low wages they can demand because of the relatively low barriers to entry for those jobs). The end result, Levy and Murnane suggest, is an unfortunate hollowing-out of the middle of the job market, with relatively well-paying jobs that required some education but were susceptible to automation dropping in number and remuneration. Many of those jobs fall into the "middle class," a critical part of society both in terms of its buying power driving the rest of the economy and its role as an aspiration for those in lower-paying jobs.

One aspect of Levy and Murnane's characterization of which jobs will be impacted by software requires examination. The implication that computers can only do jobs that use a series of rules versus those that require pattern recognition contradicts the discussion in this book's section on computer algorithms that perform pattern recognition. For some pattern recognition tasks, the computer can do a more consistent and faster job than an average person using intrinsic pattern recognition skills. For example, a computer can match fingerprints more quickly and consistently than a human. A computer can transcribe dictated medical reports with words that the average person is unfamiliar with. A computer can provide a pretty good translation of one language into another, which humans can only do if they speak both languages fairly well. Software can recognize which of millions of web pages might satisfy an inquiry in a way no individual could. Some game systems (e.g., Microsoft's Xbox with the Kinect controller) can "see" players today with a camera and mimic their movements on a screen. The camera in your mobile phone may highlight faces in an image to allow focusing on them. The list goes on, but it suggests that, to understand what jobs computers can automate, the use of the term "pattern recognition" to describe what humans do better than computers needs refinement. Increasingly, software can recognize patterns that were the sole domain of humans until relatively recently. I discussed earlier what types of tasks computer pattern recognition techniques could most easily handle—that category is growing as computers get more powerful.

Progress in computers responding to language input, perhaps one of the most difficult pattern recognition tasks, is accelerating. Mobile personal assistants that respond to spoken or typed requests in natural language have been at least partially effective, and are improving. Search engines are now accepting typed or spoken natural-language inquiries. Increasingly, "semantic" analysis is used to go beyond the specific words used and deliver what we intend.

One limit on automation using computer pattern recognition is more practical than technical. There are some things we simply want a human to be responsible for, e.g., a medical diagnosis and recommended treatment. The legal system provides some checks on automation. Tasks that might lead to lawsuits when automation goes wrong are less likely to be automated than those that are more benign.

One approach to this legal hurdle is often to make a human responsible for the result of the automation as a validator. For example, natural language processing is being used to take an unstructured medical report a doctor has dictated (e.g., a radiologist's observations from examining an x-ray image) and enter the conclusions in the fields of an Electronic Health Record (EHR). Doctors often don't have the time to enter the results into fields on a computer screen (and don't like the process). The question whether an automated system extracted the key information from an unstructured medical report can be addressed by requiring the doctor to review the result, transferring responsibility from the software to the professional. The doctor in this case will presumably find it easier to confirm the results than enter them.

The point is that conditions in which pattern recognition by software is possible cover many cases beyond what Levy and Murnane described as rule-based tasks. This observation doesn't affect their basic observation of the hollowing-out of the middle of the job market; it expands the potential hollowing.

Managerial jobs would seem the least suited for automation, but some middle-manager jobs are also affected by software advances. Today, for example, there is sophisticated software that can monitor the conversations of call center agents with customers.

The software uses speech recognition and data analytics, listening for keywords and classifying calls for managers, even detecting new trends in problems reported by customers. The software can detect if an agent is, for example, following company guidelines for specific cases by speaking a prescribed statement in a given situation. If they wish, managers can listen to specific calls the software flags and characterizes as fitting certain criteria, such as an angry customer.

Historically, the process was much more time-intensive, requiring more managers to avoid missing important cases. Managers would randomly listen to calls and act on what they heard. The new technology allows monitoring all calls, with the requirement to listen to only a few. While the random nature of the labor-intensive prior approach took luck to discover a problem that impacted the company brand and customer satisfaction, the software can be specifically designed to detect unusual patterns and flag them, avoiding the managers' listening to calls that are uneventful. While this is just one specific example, software advances that allow doing more of what previously required human skills are likely to reduce the number of jobs in middle management.

W. Brian Arthur described "The Second Economy" in a 2011 article. He noted that technological changes such as the development of railroads in the early history of the US caused a fundamental change in the economy. He argued that such a fundamental change is taking place today: "Business processes that once took place among human beings are now being executed electronically. They are taking place in an unseen domain that is strictly digital . . . I believe it is causing a revolution no less important and dramatic than that of the railroads. It is quietly creating a second economy, a digital one." He estimated roughly that "in 2025 the second economy will be as large as the 1995 physical economy." He cites as examples of this trend checking in for a flight at airports and shipping items, all of which historically have required much more human involvement, but are done today largely with computers talking to each other—computers with access to a number of information sources. Arthur characterizes this second economy as "a neural layer for the physical economy." He doesn't consider this a casual observation, but "a deep qualitative change that is bringing intelligent, automatic response to the economy.

There's no upper limit to this, no place where it has to end . . . it would be easy to underestimate the degree to which this is going to make a difference." He notes, "Physical jobs are disappearing into the second economy, and I believe this effect is dwarfing the much more publicized effect of jobs disappearing to places like India and China."

In the third quarter of 2012, US corporate earnings were $1.75 trillion, up 18.6% from the prior year; after-tax corporate profits reached their greatest percentage of Gross Domestic Product in history. At the same time, total wages fell to a record low of 43.5% of GDP. These statistics suggest the accuracy of my basic premise that companies are automating rather than rehiring employees during the current recession.

But shouldn't job loss created by automation be self-correcting at the macroeconomic level, with efficiency in the overall economy balancing loss of jobs in some industries? That seems to have historically been the case, and seems to be the assumption many economists still make. But what good does it do to improve productivity if the product, although cheaper, simply can't be afforded by enough people to justify increased production, even at a lower cost per item or service? One might think that a declining economy would provide less justification for investing in automation and that this fact would reduce automation.

But at the microeconomic level, managers in a given company see only a need to reduce costs in the face of declining revenue. The process, viewed at the level of the overall economy, can be a feedback loop that hurts rather than helps—what I referred to earlier as "what's good for the goose isn't necessarily good for the flock." The process can destroy jobs without recognition that, if all companies reduce jobs, they also in the aggregate reduce customers' ability to buy. A given company may feel that the jobs it eliminates just makes it more competitive, but that philosophy *in the overall economy* could lead to a downward spiral. Consumers must have the means to consume, and they need jobs to do so. But within any one company, it would seem irrational to react to a drop in revenue by adding jobs. Microeconomics is at odds with macroeconomics in this case.

Income trends

And there is a related long-term problem that the recession has compounded: Growth in US median income, the income at which half the population is above and half below, has declined during the first decade of this century. This is related to the hollowing out of the middle class.

The median-income family in America earned less in 2011 than it did in 1989. The strength of the economy prior to the recession hid much of this problem, as individuals borrowed more (based on high home values and easy credit) to cover the drop in income. This decline in median income is due in part to the automation of many jobs in the sector of the economy supporting the middle class.

Certainly, some cyclic behavior is normal in a recession, and classically jobs would reappear as demand recovers. But the jobs that are open require very particular skills and education that may not match those looking for work. Jeremy Rifkin's book, *The End of Work*, claimed we are entering a new phase in world history where fewer workers will be needed to produce the goods and services for the global population. "In the years ahead," Rifkin wrote, "more sophisticated software technologies are going to bring civilization ever closer to a near-workerless world. . . . Today, all . . . sectors of the economy . . . are experiencing technological displacement, forcing millions onto the unemployment roles." Coping with this displacement, he wrote, was "likely to be the single most pressing social issue of the coming century." The growth in borrowing may have hid the trend Rifkin forecast in 1995, but it is becoming more apparent now.

Robert Reich, Clinton's Secretary of Labor and a professor at UC Berkeley, in 2012's *Beyond Outrage: What has gone wrong with our economy and our democracy, and how to fix them*, expresses alarm at some current economic trends, including the increasing disparity in incomes. For example, the US Congressional Budget Office reported that income after taxes grew between 1979 and 2007 as follows:

- 275% for the top 1% of households,
- 65% for the next 19%,

- 40% for the next 60%, and
- 18% for the bottom 20%.

Reich warns that the power of wealthy individuals and of corporations is driving politics today to an unacceptable extent. For example, the Supreme Court ruled in a close decision that Political Action Committees are exercising free speech and can be used by corporations or wealthy individuals to support candidates with any amount of money they want (as long as the candidates don't control use of the funds). And lobbyists are paid to lobby for corporate interests. Many people currently in government, including those in regulatory agencies, eventually move into industry in roles supporting the companies they previously regulated from the government side. Reich argues that these interests are distorting politics and distorting the economy at the expense of all but the very rich.

Historically, there has been a basic assumption that has driven American economic progress and social stability—the economic ladder can be climbed by all, sometimes summarized as the "American Dream." The Dream assumes fairness of the system will reward talent, hard work, or innovation and allow moving up from whatever economic class one is in to at least the next. If that assumption is no longer valid, it raises huge issues for the long-term health of US society. (Comedian George Carlin has joked, "It's called the American Dream because you have to be asleep to believe it.") A decline of the middle class in size and in income can threaten that ideal.

It's obviously not just a US issue; developing countries such as China and India, particularly with their huge populations, must preserve a perception that things are getting better for all classes if their societies are to remain stable in the long run. Income inequality certainly isn't just a US problem. About 1% of households globally control nearly 40% of the world's private financial wealth. In China in 2008, 1% of households controlled more than 70% of private financial wealth, the Boston Consulting Group estimated. This income inequality and the perception of widespread corruption will be difficult for China to manage in the long run without reform. Fortunately, it appears that the current leadership recognizes this problem and is trying to deal with it.

The top 1% of income earners has a significant disadvantage that they ignore at their peril: With absolute certainty, they are outnumbered 99 to 1! I will refer to this unquestionable fact as the "1% rule."

Blind resistance to tax changes and closing loopholes that are perceived by most of the population as unfair puts the top earners and large corporations in a precarious position. Rather than fighting reform, the wealthiest should support it, lest the political pressures for that reform become irrational, leading to what might be considered punitive reforms. Irrationality has historically occurred when a substantial portion of the population is suffering and irrationality doesn't favor those who have substantial assets. Ninety-nine percent of voters can outvote one percent easily. If voting doesn't work, social instability is a threat. It's not in the interest of those with large financial assets to let irrationality develop. A careful look at history should alarm those who believe they can simply resist reform rather than looking at the health of the economy and the fair distribution of wealth as being in their interest.

I personally don't believe in the characterization of the more affluent segments of the population as villains interested only in building their own assets. The wealthiest individuals give back to society, sometimes thorough foundations they create, and hope that their legacy goes beyond simply having gotten rich. However, there is a risk that the majority of the wealthy will simply ignore the growing issue of income inequality, in part because they don't see an easy solution to support.

The 1% rule applies to corporations as well. It is easy for those running corporations to hide behind the corporate veil, insisting they are just doing what is best for the company, even as they collectively reduce disposable income that is required to buy their products. Management must understand that simply supporting positions they think will benefit them in the short term can backfire in the long run. Large companies have a responsibility to society, but it's not just good citizenship that should motivate them. They obviously have a stake in the long-term health of the full economy.

Companies, like wealthy individuals, have the choice of supporting rational change or eventually dealing with more aggressive changes in attitudes and actions. We are already seeing some of that today; some executives are being jailed for illegal activity (such as insider trading) that they may have thought was "business as usual" before the crisis. One example of a reaction to the current state of affairs would be a law forbidding "golden parachutes" for management in light of the stories about the amounts of money given executives removed after the government bailout of financial firms. The CEO of AIG, the large insurance firm that the government bailed out in the recent financial crisis, collected $250 million in 2008 when the firm collapsed, according to Robert Reich's book. With a parachute like that, there isn't much fear of being thrown out of the plane.

The supposed power of wealthy individuals and corporations over political parties should be used to help solve core economic problems and get away from slogans and from knee-jerk reactions and move to attempts to make the economy fairer. Reich characterized the problem colorfully: "Free enterprise is on trial . . . It's whether an economic system can survive when those at the top get giant rewards no matter how badly they screw up while the rest of us get screwed no matter how hard we work." A wide adoption of that point of view could lead to a reaction that strikes at the fundamental protections allowing one to build and hold wealth.

What does this all have to do with software? The point this book argues is that the continuing productivity improvements created by software are accelerating, reducing some job categories that historically supported a large middle class at a rate that doesn't allow the normal adjustment to creative destruction. It will take innovative solutions (some of which I recommend later in this book) to address these issues, not attitudes driven by partisan slogans or obstructionist behavior by the wealthy or corporations. The deep reforms required can't happen without our working together. To the degree that corporations and wealthy individuals don't support reform, it may be that they don't see any reforms that they think make sense. I suggest one in this book that I believe will work to create jobs, including jobs that should raise median income, so this admonition comes with a call for specific action. I devote a long discussion to this alternative in a later chapter.

W. Brian Arthur suggests that "the main challenge of the economy is shifting from *producing* prosperity to *distributing* prosperity." He suggests the possibility of new job categories created by the digital economy or moving to shorter workweeks.

I'm uncomfortable with an approach such as shorter workweeks that simply increases the number of workers without increasing the amount of work done. Longer *paid* vacations or salary *increases* that maintain income levels with a shorter workweek *decrease productivity* and national competitiveness. Unpaid vacations or lower income because of a shorter workweek simply continue the decline in median income and are not likely to generate much economic activity.

The full solution isn't likely to be short-term. Changes in education, for example, starting at the elementary school level, are critical to improving the long-term skills of the workforce, but aren't easy to implement quickly. But investment in areas such as education can create new and better jobs immediately. Teachers are a critical element in any successful society. An earlier chapter suggested technology could make teaching a more attractive profession.

Is there a way that continued software evolution could help reverse the trend toward automation taking jobs? Possibly. As we connect more tightly with our technology, it can help us do more as individuals. The human-computer connection makes a powerful team, and companies can exploit this flexible combination to make automation expand human jobs rather than eliminate them. Human-computer partnerships might cause some machines to lose their jobs! This point is discussed in more detail in a later chapter.

Job trends

Let's take a brief look at some of the job categories that are growing and shrinking as we recover from the 2008 recession. Table 1 is a snapshot of projections of which jobs will grow rapidly (29% or more) by the US Bureau of Labor Statistics as of October 2012. Table 2 is the agency's projections of jobs that are likely to decline. Please note that these are the two extremes in the source data, which also includes cases (not listed here) where jobs are expected to grow, but less than 29%.

In Table 1, there are many healthcare jobs, from doctors to physical therapy assistants, that require at least specific job training and experience. Some, however, don't require a bachelor's degree, and offer an opportunity for those willing to get the specialized training, since the rapid growth will create demand. Since going back to school looks better on a resume than unemployment, this is an alternative for the unemployed who have some resources that let them defer working. Government policy can help here with student loans, although the growth in student loans and some collection problems in the program have put pressure on this option. The downside of back-to-school is that it doesn't create immediate employment, and students are not likely to have much disposable income. Nevertheless, it would ease shortages that might otherwise occur in the future, reducing a mismatch between the labor force and available jobs.

Another category in Table 1 that stands out is jobs that require a high school diploma or less, but require skills typically developed on the job, e.g., brickmasons, helpers for electricians, carpenters, and plumbers, and glaziers. This reflects in part the current rebound in construction from the low of the recession.

A characteristic of most jobs in Table 1 is that they are hard to automate, although I would argue that some, such as interpreters and translators, can at least be partially automated by today's evolving software (and there may not be the growth the Bureau of Labor Statistics expects). But many of the jobs require the dexterity and mobility of the human body, as well as human judgment and creativity. Where the issue is training, the human-computer connection could ease that requirement, e.g., by providing portable computer guidance in doing jobs such as medical equipment repair and healthcare assistance.

The declining jobs listed in Table 2 largely reflect areas of consumer spending that the recession has hit, such as restaurants and manufacturing. The good news is that these are still important parts of the economy, and a general recovery in the economy may help restore these jobs, despite the prediction of decline. Most of the declining jobs would be difficult to automate, so the good news is that automation is unlikely to eliminate the jobs. But the bad news is that, if the economy continues to stagnate, these job categories will continue to suffer.

Table 1: Most rapidly growing jobs (10,000-49,000 new jobs, 29%
growth rate or higher projected), Source: US Bureau of Labor
Statistics Occupational Outlook Handbook (as of October 2012,
www.bls.gov/ooh/occupation-finder)

Occupation	Education	2010 Median pay
Medical Scientists, Except Epidemiologists	Doctoral or professional degree	$75,000 or more
Optometrists	Doctoral or professional degree	$75,000 or more
Marriage and Family Therapists	Master's degree	$35,000 to $54,999
Mental Health Counselors	Master's degree	$35,000 to $54,999
Occupational Therapists	Master's degree	$55,000 to $74,999
Physician Assistants	Master's degree	$75,000 or more
Interpreters and Translators	Bachelor's degree	$35,000 to $54,999
Health Educators	Bachelor's degree	$35,000 to $54,999
Meeting, Convention, and Event Planners	Bachelor's degree	$35,000 to $54,999
Mental Health and Substance Abuse Social Workers	Bachelor's degree	$35,000 to $54,999
Database Administrators	Bachelor's degree	$55,000 to $74,999
Medical Equipment Repairers	Associate's degree	$35,000 to $54,999
Cardiovascular Technologists and Technicians	Associate's degree	$35,000 to $54,999
Occupational Therapy Assistants	Associate's degree	$35,000 to $54,999
Physical Therapist Assistants	Associate's degree	$35,000 to $54,999
Diagnostic Medical Sonographers	Associate's degree	$55,000 to $74,999
Brickmasons and Blockmasons	High school diploma or equivalent	$35,000 to $54,999
Glaziers	High school diploma or equivalent	$35,000 to $54,999
Opticians, Dispensing	High school diploma or equivalent	$25,000 to $34,999
Physical Therapist Aides	High school diploma or equivalent	Less than $25,000
Helpers—Electricians	High school diploma or equivalent	$25,000 to $34,999
Helpers—Pipelayers, Plumbers, Pipefitters, and Steamfitters	High school diploma or equivalent	$25,000 to $34,999
Cargo and Freight Agents	High school diploma or equivalent	$35,000 to $54,999
Helpers—Brickmasons, Blockmasons, Stonemasons, and Tile and Marble Setters	Less than high school	$25,000 to $34,999
Helpers—Carpenters	Less than high school	$25,000 to $34,999

Table 2: Declining jobs and negative growth rate, Source: US Bureau of Labor Statistics Occupational Outlook Handbook (as of October 2012, **www.bls.gov/ooh/occupation-finder**)

Occupation	Education	2010 Median pay
Air Traffic Controllers	Associate's degree	$75,000 or more
Aerospace Engineering and Operations Technicians	Associate's degree	$55,000 to $74,999
Drafters, All Other	Associate's degree	$35,000 to $54,999
Forest and Conservation Technicians	Associate's degree	$25,000 to $34,999
Desktop Publishers	Associate's degree	$35,000 to $54,999
Jewelers and Precious Stone and Metal Workers	High school diploma	$35,000 to $54,999
Drilling and Boring Machine Tool Setters, Operators, and Tenders, Metal and Plastic	High school diploma	$25,000 to $34,999
Fiberglass Laminators and Fabricators	High school diploma	$25,000 to $34,999
Lathe and Turning Machine Tool Setters, Operators, and Tenders, Metal and Plastic	High school diploma	$25,000 to $34,999
Chefs and Head Cooks	High school diploma	$35,000 to $54,999
Farmers, Ranchers, and Other Agricultural Managers	High school diploma	$55,000 to $74,999
Food Service Managers	High school diploma	$35,000 to $54,999
Animal Breeders	High school diploma	$25,000 to $34,999
Coil Winders, Tapers, and Finishers	High school diploma	$25,000 to $34,999
Correspondence Clerks	High school diploma	$25,000 to $34,999
Electrical and Electronic Equipment Assemblers	High school diploma	$25,000 to $34,999
File Clerks	High school diploma	$25,000 to $34,999
Floral Designers	High school diploma	Less than $25,000
Gaming Cage Workers	High school diploma	$25,000 to $34,999
Loan Interviewers and Clerks	High school diploma	$25,000 to $34,999
Insurance Appraisers, Auto Damage	Postsecondary non-degree award	$55,000 to $74,999
Fishers and Related Fishing Workers	Less than high school	$25,000 to $34,999
Cooks, Fast Food	Less than high school	Less than $25,000
Food Preparation and Serving Related Workers, All Other	Less than high school	Less than $25,000
Loading Machine Operators, Underground Mining	Less than high school	$35,000 to $54,999

Some of the categories declining are manufacturing jobs, despite the fact that this data reflects a period of slow recovery, perhaps suggesting further that jobs are being replaced by automation where they can be. None of the fast-growing job categories are in manufacturing. The issue then returns to the question of whether there is an intrinsic issue in automation reducing jobs beyond a critical mass to sustain the economy, and how we can address this issue.

Another message we could take from these projected trends is that in general the growing job categories require more education than the declining job categories. This indicates that investing in education—a topic that this book has discussed—can help fill jobs that are available. The median salary of the jobs appears to be correlated with education, and probably helps account for the higher median salary in the growth category (Table 1). This is of course just a partial snapshot of job trends at a particular stage of recovery.

14

Human Connection to Computers as Part of the Solution

Our technologies help define what we do every day, and that definition changes over time. The acceleration of that change through software has an important impact. This section is an attempt to understand to what degree our increasing connection with computers can increase our human capabilities in a work environment and to what degree it can compensate for the productivity enhancements that are reducing many current categories of jobs.

A barcode scanner makes the identification of a product available through software. A human could not easily learn to read a barcode. But a human is required to find the barcode and scan the barcode, a good example of the flexibility and mobility of a human being required to use a technology. The partnership allows software to produce information used immediately, such as price at a checkout counter, and information that is used more strategically, such as maintaining an inventory count.

A spelling checker in a word processing program or the use of a barcode scanner are cases where the benefits of technology are more tightly coupled to an individual than technologies that tend to operate independently, like a heating system. I may change the thermostat on a heating system occasionally, but it is not as intimately tied to my thinking activities as a spelling checker.

Humans are amazing biological "technology." We are mobile in ways that technology struggles to match, avoiding obstacles, walking, climbing, and much more. We are flexible in using our appendages and senses to accomplish a task when unexpected variations occur.

We can put together our observations and knowledge to recognize patterns and connections with insights that come from a life of experience. We understand other people and how they are likely to react in a given situation (at least to some degree).

How far can this tighter coupling of people and computers go, and to what degree can computer intelligence couple with human mental strengths in combining computer intelligence with human strengths in mobility, dexterity, intuition, and adaptivity to create an end result that neither humans or machines can do alone? Just to take a simple example of this synergy, a spelling and grammar checker in a word processing package helps provide a job for someone who creates content well, but has a problem with the details, but the spelling checker isn't going to create a good newspaper article by itself.

I'll highlight the possibilities for this synergy in these categories:

- Supplementing human capabilities in the workplace;
- Allowing work from home or distributed locations;
- Computer training teaching us skills or knowledge more efficiently;
- Making it easier to build businesses in smaller markets;
- Enabling new methods for creative work and means to start a business;
- Aiding workers with disabilities; and
- Automating tasks through a human-computer partnership, rather than by a computer alone.

I intend these particular areas to be examples, and not a definitive list of all possibilities. I will suggest in the next chapter a way that we can motivate companies to be creative in finding jobs for people through the human-computer connection (or other ways) rather than turning to full automation as a first preference.

The human-computer connection in the workplace

Software has long been part of the work environment. Some software, for example, improves the efficiency of communications between employees, email being an obvious case. Other productivity tools, such as spreadsheets, word processors, and presentation tools are often part of a job as well.

Most companies have software programs that are fundamental to their operation. Database software stores information on customers, pricing, orders, inventory—the basic data required to run a business.

There is software tailored to supporting a large sales force, allowing salesmen to keep track of prospects and actions required to close a sale, as well as reporting a sale. The evolution of sales software is an example of the tightening human-computer connection. For example, increasingly, sales software has a mobile component, so that entries can be made from the field. For Saleforce.com's software, for example, there are outside options that let entries be simply spoken into a phone to enter data into the database, allowing a salesman to update status while the information is fresh, perhaps driving from a meeting.

Combining human and software technology can create job categories not possible (or at least not as effective) as either capability alone. Warehouse workers that drive carts through distribution centers picking products for an order could in theory be replaced by some sort of automated conveyer belt moving past bins that pushed products onto a conveyer belt, and some products might be amenable to such systems. But a warehouse with a wide variety of products of varying sizes and shapes would run into many problems with items that didn't get pushed properly, multiple items being pushed instead of one, assumptions that a push occurred when it didn't (perhaps because a bin was empty), possible damage if a product is pushed on top of another, etc.

Humans are very good at adapting to this variety of objects and conditions in a warehouse, and most warehouses use people to fill orders. Where workers have a disadvantage is in connecting with the computer software holding the orders and keeping track of inventory, software that is usually called the Warehouse Management System (WMS). Reading a screen and entering results manually would significantly slow the process since the worker is handling many of the items and navigating a vehicle.

The problem has been solved by giving the workers portable wireless devices where they report that an item has been picked up for an order, that a bin is empty, and other information the WMS needs. Increasingly, these systems use voice interaction, so that hands and eyes are left free. The system speaks the next item to be picked and the warehouse bin where it can be found in a synthetic voice. The worker says what she has just done when picking the item, typically scanning a barcode and having the computer confirm that it is the right item, or noting that the item is out of stock.

This "hands-free" operation using speech recognition software is not the exception, but, perhaps surprisingly, very widely used in distribution centers. (The same system has been adapted for some healthcare environments where sanitary requirements motivate hands-free reporting of activities performed.) Most of these systems allow voice interaction in several languages and use technology that can adapt to heavy accents. These voice systems thus allow a diverse group of workers to interact with the computer system.

To show the effectiveness of this human-computer partnership, I'll quote some numbers from a case provided by Vocollect, a provider of voice systems for distribution centers. One of their customers, Fox Racing, makes specialized sports clothing. To fill customer orders, Fox Racing shifted from a system using minimal automation to a system where warehouse workers wear headsets. The workers converse hands- and eyes-free with a speech recognition and speech synthesis system that gives them directions and to which they respond by confirming task completion. With the voice system connecting the worker directly to the computer system rather than the previous manual approach, picking of an order was 50% faster, accuracy increased from 82% to a remarkable 99.9999%, and training time went down from one day to 1.5 hours, with workers reaching the top levels of productivity after only one day on the job. We've mentioned that US workers are more productive than many other countries, and such automation is an example of how that is achieved.

Today, one aspect of efficiency at work is the use of personal smartphones, laptops, and tablet computers, the "Bring Your Own Device" (BYOD) trend. Studies by market research company IDG Connect in late 2012 found that 60% of the US IT professionals surveyed bring their own device to work. A near unanimous 95% believe it positively improves their work/life balance, and 73% think it enhances their colleagues' productivity.

Some companies provide technical support for personal smartphones, tablets, and laptops used at work. A Gartner survey in late 2011 found that, of 938 businesses surveyed in nine countries, 32% said they support personal smartphones, while 37% said they support tablets. The support for personal devices was 44% in the BRIC fast-developing countries (Brazil, Russia, India, and China), which have a larger number of young workers.

A Cisco study conducted in Spring 2012 also found that BYOD's growth is not a U.S.-only phenomenon nor limited to large companies. Globally, 89% of IT leaders from both large enterprises and midsize companies support BYOD in some form. And 69% view BYOD "somewhat" or "extremely" positively.

A separate survey of young workers (college-educated employees between the ages of 20 and 29) in fifteen countries in 2012 by research firm Vision Critical found that more than half viewed it as their "right" to use their own mobile devices at work, rather than it being a "privilege" granted by the employer. The survey also indicated that a large percentage in this group would basically ignore even a limited company policy against the use of the devices. Of course, the enthusiasm for mobile devices goes well beyond younger workers. Chief Information Officers cite the common case of the CEO asking that his favorite device be supported by IT.

A realistic attitude is to make the most of the good aspects of BYOD. The accelerating progress in coupling human and computer intelligence will make these benefits even more pronounced than those already being cited by enthusiasts. Most companies seem to be moving in this direction. A November 2011 survey of 605 C-level executives by Avanade, a managed services provider, found that 73% of respondents said increasing the use of employee-owned technology is a top priority in their organization, and 60% said they are now adapting IT infrastructure to accommodate personal devices rather than adopting policies that restrict their usage.

The Avanade survey respondents said the main attraction of BYOD is increasing productivity. Most (58%) of respondents said the greatest benefit of BYOD is the ability to allow employees to work from anywhere, while 42% said the greatest benefit is that employees are much more willing to work after hours. Another survey conducted in August 2012 by Vanson Bourne, a research organization, for the company Citrix concluded that the total number of organizations that have implemented "mobile workstyles" will rise from 24% in 2012 to 83% in 2014. The survey found that only 9% of the organizations do not have plans to adopt mobile workstyles.

These findings suggest that using computers as a constant enhancement to our human abilities is already extending to workplaces well outside the warehouse. Once these mobile devices are available to us at work as well as outside work, we automatically benefit from any enhancements to their user interfaces, software, and services.

I've emphasized mobile devices because their capabilities can be enhanced by both on-device apps for specific business applications and cloud-based applications connected to the devices. Companies can use the general-purpose nature of these devices and software development tools supporting these devices to create company-specific solutions that can expand jobs and make them more productive. Of course, more conventional extensions of software on PCs and elsewhere will continue to provide us with resources that expand human capabilities.

The distributed workplace

Another area where computer technology expands job possibilities is through networking, particularly the Internet, where an individual can work remotely from the workplace, normally at home. Mobile devices provide the worker even more flexibility in where they work, but even if the worker is tied to a personal computer at home, they can work almost as if in the office, interacting with other employees by email, phone calls, and even videoconferences if needed.

If the worker works only from home and uses their own computer equipment, there are obvious cost advantages to the company. It may be an advantage both to the company and the worker if this allows more flexible work hours, e.g., handling technical support calls in the evening. The worker avoids the cost and time lost in commuting and can potentially work flexible hours. The last point could be particularly important, for example, to a parent of a young child. The overall efficiency could mean that even a smaller salary results in more net income after expenses for the worker. This type of work is best for task-based jobs, where the worker's contribution is measured by specific accomplishments, rather than time spent, eliminating issues of it being harder to monitor the individual's time on the job. (Realistically, if a time clock is the only measure of an employee's contribution, the company is in trouble.)

Continuing education

Learning new skills to match the changing job environment is an option. Software makes it easier to continue one's education through online courses and webinars. Local colleges typically have extension courses that often use the web as an information resource or for completing assignments. These resources allow individuals to learn new skills more quickly and inexpensively. The trend also provides options to people who need to keep their skills updated or learn new skills to be fully effective in their jobs. Instructors and course creators often get feedback through these web resources on where courseware needs to be improved. Lynda Gratton, director of the Future of Work Consortium and author of *The Shift*, notes that online education platforms like MIT's OpenCourseWare, Open Yale, iTunes, U, and Khan Academy enable students worldwide to have very similar learning experiences and work towards similar qualifications.

Making it easier to start a small business

Small businesses have historically been important generators of jobs (and often they turn into larger businesses). Some of these businesses simply are jobs for the owner, without a payroll. About three quarters of all US business firms have no payroll. Most are self-employed persons operating unincorporated businesses, and may or may not be the owner's principal source of income. Such businesses don't generate huge revenues; the US Census Bureau estimates that these "non-employers" account for only about 3.4% of business receipts. But considering that there are some 22 million such firms in the US generating about $1 trillion in sales annually, they are obviously an option for many people, and, of course, those people spend money that provides jobs for others. The Census Bureau estimates that these non-employee companies are responsible for a similar number of "establishments," implying that being self-employed doesn't mean you don't pay rent and utility bills, among other expenses such as advertising that contribute to economic activity.

Technology makes it increasingly easy for such independents to operate and reach customers. Some of the technology described in the "distributed workplace" section applies to this category of workers, as well as many tools and services available through the Internet, including ways to reach customers.

While the number of non-employee firms is large, small business employers with at least one employee have a much larger impact on the economy. There are about 6 million businesses with less than 500 employees each, employing some 60 million people. Today, software and Web-based resources make it easier and less capital-intensive to launch and grow such businesses. The most obvious symptom of this trend is the growth in "cloud" services. Small businesses can "rent" much of the software (and the hardware supporting it) that they would have to buy and manage otherwise.

Creativity

Many people have creative talents and interests, but find it difficult to translate those into an income-producing activity. They often treat such interests as a hobby or abandon the activity entirely under the pressure of jobs and other responsibilities.

Technology is providing increasing outlets for such talent through routes such as YouTube, self-publishing, and tools for creating hardware prototypes or small quantities of customized products. Genevieve Shore, the chief information officer and director of digital strategy for the publisher Pearson, stated in an online article in October 2012 that the UK's creative industries constitute one of the fastest-growing sectors in the UK, contribute 6% of gross domestic product (GDP), and employ more than two million people, largely because of the increased outlets for creativity presented by the Web. Shore said that the value of UK digital fiction sales in the first half of 2012 was up 188% over the same period in 2011. And specialty web sites provide a route to selling creative products that one might otherwise consider too narrow a market.

Chris Anderson in *Makers* points out manufacturing innovations that allow making customized objects in small quantities, e.g., cups with personal slogans or images. These newly evolving options open a market for artisans to create products that can be sold in small quantities to individuals who want something tailored to their interest rather than something mass-produced. Making such products can be a business rather than a hobby because of the ability to reach customers searching for such products on the Web. Further, custom products can be sold with a "uniqueness premium" in the price.

Software applications, particularly for mobile phones, have been democratized in the sense that one can create a specialized application and sell it through an App Store without major capital or marketing expenses. The number of apps available for the major mobile operating systems is in the hundreds of thousands, a market that didn't exist a few years ago. Apple has pointed out that its iOS devices support an app development community, adding "more than 210,000 iOS jobs to the US economy since the introduction of iPhone in 2007"; Apple indicated that it had paid more than $4 billion in royalties to developers through the App Store midway through 2012.

Can this trend impact the economy in a significant way? For the people it enables, it can be both self-fulfilling and generate some income. Even as a part-time job, these opportunities can generate increased disposable income, helping the economy in general.

Accessibility

Some jobs go unfilled because of a shortage of people with particular skills, such as technical backgrounds or simply experience in a business area. In some cases, those people are available, but have developed disabilities that are presumed to limit their ability to work. For example, a person may have developed carpal tunnel syndrome, perhaps from overuse of a keyboard because of extensive programming or writing, have lost their vision or hearing, or become physically disabled.

Computer technology makes it possible for people with disabilities to do work that would otherwise be difficult for them. Software that helps them use computers is often labeled as "accessibility" tools. For example, these tools allow the blind or visually impaired to have text on the screen spoken to them. Individuals with manual impairments can dictate text and control much of computer behavior today by voice, minimizing or eliminating use of the keyboard. As an example of industry effort in this regard, the World Wide Web Consortium (W3C) has a Web Accessibility Initiative (www.w3.org/WAI) that provides guides to accessibility resources.

The human-computer connection and jobs

The increasing connection between humans and computers creates a combination that can do jobs more effectively than either alone. This is not new, as I have noted, in that tools in all their varieties have always made humans more effective at certain jobs. Many of those tools, e.g., a hammer or shovel, have augmented human manual dexterity. Others, such as spreadsheet software, have augmented human mental capabilities.

As computers more tightly integrate with our *language* capabilities, and thus with our core reasoning and thinking abilities, this trend of increasing access to computer intelligence will continue and accelerate. The earlier example of warehouse stock picking augmented by voice interaction with computers is an example of the efficiencies that this connection can create.

Every industry and job category has needs that are particular to that industry or category. Every company knows where it has bottlenecks or where problems arise. As the human-computer connection evolves, it will be possible to address more of these problem areas with a human-computer solution. In many cases, a human coupled to computer intelligence can be more effective than a pure software solution or a pure human solution, and in many cases the combination will be the only way to solve the problem. The "package" of flexibility, mobility, intuition, and adaptability in humans is unmatched, despite the more adventurous speculations of futurists or science fiction writers.

In some cases, the human-computer solutions will be internally developed solutions. In others, companies will see business opportunities in creating and selling applications that apply to many companies, such as the warehouse or medical reporting solutions previously discussed.

The result may simply be better quality of product or performance. In other cases, one worker may be able to do more work, reducing the number of workers necessary. In that case, you might ask, is there really a difference between this type of automation and automation that doesn't involve human participation? In either case, you might argue, the number of jobs is reduced. A significant difference, however, is that a human earning a wage is involved. Presumably, if that human is doing a job that originally required more than one human, there is room to pay them more and still come out ahead. Increasing disposable income creates consumption—all else being equal—and creates jobs indirectly. This type of efficiency is how productivity can improve median income of those working. There is one less job lost compared to full automation.

I could make this book much longer by going industry by industry and trying to suggest where the human-computer connection could create synergy. But there are obviously people in those industries that will come up with better solutions than I could imagine. In the next chapter, I suggest a particular government policy that will motivate companies to look for this type of solution.

There will of course be jobs created by a recovering economy that have nothing to do with the tightening human-computer connection. But the key is to get the economy on a track that creates new jobs, and the incremental jobs created by the human-computer connection will be a substantial contributor.

15

An Automation Tax

To summarize, productivity has its limits. Automation can reduce personnel costs for companies, but can destroy jobs at too fast a pace for the economy to absorb. Increased productivity can reduce the price of goods, but someone still has to buy those goods. If unemployment remains stubbornly high or even grows, at some point the loss of jobs will overcome the advantage of lower prices, and a long-term recession could ensue.

Monetary policy—making money more available for loans or capital equipment—can't cure this structural problem in the economy. If taken too far, adding to the money supply will only create another bubble to burst. Ironically, the low cost of borrowing can motivate more capital expenditures directed at automation, and reduce jobs. Lower cost of money can make it even cheaper to "hire" computers rather than people.

The companies benefiting the most from productivity with the highest revenue per employee—also tend to be very profitable and generate stockpiles of cash, cash that doesn't contribute much to the economy. To compound that effect, US companies generating profits abroad are not repatriating the resulting cash; they want to avoid the resulting taxes.

I've outlined some ways that software can augment what humans can do and what machines can do by combining the strengths of both, extending a trend that has been important historically. However, motivating companies to take advantage of this opportunity or to find other ways to create jobs is a challenge. We need a counterbalance against the perceived advantage of "hiring computers" instead of hiring people. This section thus suggests an automation tax as a mechanism to harness the creativity of companies in this cause by providing a financial incentive for companies to come up with solutions that create jobs. In short, the automation tax is a penalty for preferring automation to people when there is a fair tradeoff.

I prefer a solution using taxes because it has the leverage of the feedback principle. It encourages certain behavior without requiring it and has minimal regulatory requirements. The specific solution I suggest—an "automation tax"—is my preference, but other approaches are described in later sections.

What is an automation tax?

If software is to take over many jobs, why not have an income tax on software? We could perhaps think of it as a payroll tax on computers. As put, this may sound like a joke, but an automation tax is a meaningful possibility.

Companies unfortunately see a disadvantage in hiring people over hiring computers. People come with baggage such as payroll taxes and employee benefits. They can file complaints about unfair treatment. They require vacations and sick leave. In 2014, many companies that don't do so now will be required to provide employees health insurance at a prescribed minimum level.

In contrast, automation basically requires more software and computers. Using that software for automation is increasingly easier as software gets more features, is more reliable, and as the cost of a given amount of computing drops every year. Computers essentially come with a guaranteed salary reduction over time; humans generally require increasing compensation over time.

An automation tax described as a payroll tax on computers conveys the basic concept. It helps level the playing field. The automation tax serves two purposes: (1) it provides an incentive for a company to create jobs by means such as investing in human-computer synergy; and (2) it provides government revenues that, properly used, can create more consumption and thus boost the economy.

Companies pay this tax, not individuals. This is perhaps appropriate in more general terms of asking companies to pay their fair share of government taxes. The Center on Budget and Policy Priorities reported in 2011 that corporate tax revenues as a share of GDP was near historic lows, due to the many deductions possible (despite a high corporate tax rate).

The goal suggested by the analogy to a payroll tax on computers can be accomplished in a practical way. I propose that a national automation tax be based on the *ratio of a company's revenues (total sales) to their number of employees*. The tax would be more if this ratio is higher. The basic concept is that a company that achieves high revenues with a small number of employees is using automation to do so, while one with more employees to achieve the same revenue is using less automation. A company using more automation would pay a higher automation tax than a company using less automation.

One could use the ratio in the other direction—employees per million dollars in revenue, for example—if that were deemed more intuitive. In that case, one would of course target increasing that ratio. I'll use revenues per employee to be consistent.

According to an Apple web site, Apple had about 47,000 employees in the US and 70,000 worldwide in 2012. The company was running at a revenue rate of about $40 billion per quarter. With about the same revenue, GM had about 70,000 employees in the US and about 200,000 worldwide. The ratio of worldwide revenues to employees at Apple is about $571,000 and at GM is about $200,000 per worldwide employee. Looking at the reverse ratio, Apple has 1.8 employees per million dollars of revenue worldwide and GM 5.0 employees per million dollars, almost three times as many.

Companies with high values of this revenue-to-employee ratio will probably argue that they create jobs outside the company and this compensates for their low internal head count relative to sales. The Apple web site in 2012, for example, claimed that 257,000 jobs were supported indirectly by Apple at other companies, in fields that included the development and manufacturing of components, materials, and equipment; professional, scientific, and technical services; consumer sales; transportation; business sales; and healthcare. GM doesn't provide similar numbers, but certainly there are a huge number of jobs selling and servicing cars, and perhaps one should include gas stations, tire manufacturers, etc. While creating outside jobs is relevant, it isn't something that can be easily measured in a consistent way or easily validated. Some of the jobs Apple refers to are outsourced manufacturing jobs in China, for example, not an ideal substitute for US jobs, at least from a US perspective.

An automation tax could encourage companies to evaluate and take a more balanced view of their cost of automation versus the cost of retaining or adding employees. Employees are often viewed as having significant overhead costs, such as the management employees required, payroll taxes, vacation time, healthcare insurance, retirement plans, complaints, etc. But automation has costs that might not be fully evaluated without a comparison to retaining, hiring, or retraining workers. Computers and software also must be managed; that's why most companies have an Information Technology (IT) group. Computers can be hacked, and thus cause security and public relations problems. Computers don't always do the job they were intended to do; when that happens, it's called a "bug," and the software doesn't even get a reprimand, simply a lot of attention. And software must be purchased or developed, so there is a cost beyond ongoing maintenance. The automation tax is an attempt to make the cost of automation more obvious and better balance the tradeoff between humans and machines.

The automation tax doesn't prescribe specific job targets or specific job categories. It allows companies to optimize the jobs they create without the government trying to mandate specific goals. A company can simply pay the tax, if that is the optimal approach for that company. The feedback principle is at work here; the government provides an incentive for companies to do something that benefits the entire economy (including the companies), but simply pushes in the right direction without specifying details. The tax code has always been used to promote, but not mandate, certain behaviors. The mortgage deduction promotes home ownership, but doesn't mandate it. Tax-free growth of retirement accounts promotes retirement savings. And the list goes on. Even if all companies decided just to pay the tax rather than try to increase jobs, the automation tax can be spent by government in ways that promote employment or increase disposable income, as will be discussed in a later chapter, providing increased impact on job growth.

Returning to the payroll tax analogy, companies that hire fewer people pay fewer payroll taxes. The payroll tax in the US helps fund social security, Medicare, and unemployment insurance. In Europe, payroll taxes are even higher than in the US. The safety net that these government programs provide to a company's employees reduces the need for such programs to be financed by the company (e.g., more expensive retirement plans). And companies sell their products to retirees whose purchases are made possible by social security or other safety net programs. Companies who pay less payroll taxes as a fraction of revenue (because they are more productive due to automation) don't pay their fair share of these societal necessities. The automation tax reduces that inequity.

The details

The ratio of revenues-to-employees driving the automation tax rate is a summary of my proposal. There are important details.

What numbers are used in the ratio of revenues to employees? I recommend using revenues generated *within* the taxing country and employees *within* the country in the ratio. The revenues within country are generated from citizens of that country, and seem appropriate to use in calculating a tax to be collected by that country. And the employees within country are the jobs a government wants to encourage.

If calculated based on in-country employees, the automation tax has the further benefit of penalizing outsourcing of work to other countries, whether with their own employees out of country or simply using companies in other countries to produce products. Companies can reduce the automation tax by repatriating some jobs. Perhaps the automation tax should be called an Outsourcing and Automation Tax, with the benefit of a nice three-letter acronym—OAT. But my focus is on countering over-automation, so I won't emphasize this aspect of the tax.

Another detail is the number-of-employees calculation. A company could artificially increase the number of employees by hiring part-time employees versus full time employees to do the same work. This can be avoided by using employee hours in the calculation rather than number of employees, i.e., the ratio of in-country revenue to in-country employee-hours. (I will continue for brevity to use simply "employee.")

To make sense, the automation tax should increase as a percentage as the revenue-per-employee grows, making it more attractive to create jobs than to replace them with automation. But what should that percentage be applied to, income (profits) or revenues (sales)? Applying the tax to revenues makes it more like a national sales tax. Applying it to income means that a company that is less profitable would pay less than one having the same ratio and doing better.

I prefer applying the tax percentage to revenues, as suggested earlier. Profits can be manipulated with deductions and other accounting complexities much more than revenues. If the tax is applied only to in-country activities, it is much easier to calculate in-country revenues than in-country profits.

If applying the automation tax to profits were preferable for political or other reasons, I recommend it be applied as a percentage increase or decrease of taxes that would be paid otherwise, making it either a tax penalty or a deduction depending on the revenue-to-employee ratio. This approach could avoid the possible need to reduce other corporate tax rates, a point to which I will return.

Another decision is the overall size of the tax. The tax on an individual company must be enough of a tax penalty for a high revenue-employee ratio that it provides a meaningful incentive for the company to create jobs, or it will have no significant impact. But if it increases company taxes nationwide too much, it could stunt investment and overall growth.

The automation tax decreases as a company with given revenues hires more employees, but what amount of tax should be generated nationally and how should that be reflected in the percentages in the tax formula applied to each company? One approach to setting the level is to look at overall tax income from companies nationwide and set the level of the automation tax so that it would increase overall taxes a substantial amount, say 20%, based on a snapshot of the revenues and number of employees of companies to which the tax would apply. The percent increase should be high enough to have a meaningful impact on corporate decisions.

If the total increased tax burden is deemed too high using this approach, the standard corporate tax rate (excluding the automation tax) could be lowered overall to reduce the total increase in government tax revenues. If the basic corporate tax rate were reduced, some companies with low revenue per employee would pay lower overall taxes despite the automation tax. As noted, this same objective could be accomplished by applying the automation tax as a tax increase or deduction that depended on the revenue-employee ratio.

With an overall target nationwide for the tax, e.g., the 20% mentioned, the automation tax rate for a given revenue-to-employee ratio required to achieve this goal can be calculated from statistics on the ratio and revenues across companies. The rates would be chosen so that they generated this net result, given a snapshot of current data.

Suppose, over time, the tax achieved its goal and companies on average improved this ratio (hired more employees per dollar of revenue). The automation tax has a built-in reduction mechanism based on its success.

There are other details that would be addressed in a formal law creating the tax. For example, the smallest businesses should be excluded to avoid penalizing an organization that is efficient because a few employees working long hours are the source of its productivity.

Another tax?!

The usual view of taxes is that they take money away from individuals and corporations, resulting in less spending and investment, and hurting economic expansion. Many would argue that, when increased taxes are spent on "stimulus" spending, it is usually inefficient. "Stimulus" often becomes "subsidy," as it becomes difficult to remove the stimulus without hurting the economy or at least a vocal segment of voters. The model of increasing taxes solely to fund stimulus spending is suspect for another reason—it smacks of centralized planning since the spending makes assumptions about the economy based on someone's model. Even if there were a full understanding of how the economy works, government spending is usually some political compromise that no one fully likes.

The automation tax is not a tax-spend plan. Even if all the receipts from the tax went to reducing the deficit, the tax encourages behavior by companies that will increase jobs and build the economy (and other taxes) through creating more disposable income in the economy. *The stimulus is built into the tax*, not generated through its receipts.

If the automation tax is applied as an adjustment to existing taxes, it can be calculated, as previously noted, as an increase or decrease in those taxes. In that case, it will be a deduction rather than a tax for some companies.

Corporations might instinctively fight a corporate tax. If they look at the economic reality, however, they might realize that they must be part of the solution to maintaining an economy that won't eventually reduce revenues and kill profits. Corporations face more draconian consequences than a tax if the economy collapses. There is already resentment toward high corporate executive salaries and bonuses.

The issue goes beyond an individual corporate focus. Recall Acemoglu and Robinson's argument that, when an elite overprotected the status pro, nations failed. Perhaps this is an overdramatic characterization of the current situation, but it is certainly not unreasonable to expect company management to be concerned about the health of the political and economic system in which they must operate.

Even the Chinese leadership is showing signs of recognizing the danger of inequality of income. The 99% or more Chinese who are not part of the leadership constitute so many people that complete repression is impossible in the long run. Protests in China against corruption and even American-style protests against building manufacturing plants that pollute have been allowed with less intervention than in the past. The very fact we know about this is a symptom that the government is not preventing news of such demonstrations from travelling the Internet inside and outside China.

"It's the economy, stupid" has been a staple of political campaigns since James Carville made it a basic tenet of Bill Clinton's first campaign. It certainly was a basic theme in the Obama versus Romney campaigns, with even the debate on foreign policy bringing up economic issues. We have an intrinsic economic problem, not only in the US, but worldwide.

As politicians review options for improving the economy, it should become obvious that the automation issue must be addressed, and an automation tax may provide a simple approach, simple through using the feedback principle rather than an attempt at detailed engineering of the economy.

Supporting the economy with monetary policy is a short-term crutch. I'm concerned that the very low interest rates we are seeing today may be creating their own bubble. Eventually, interest rates must rise. When they begin to do so, financial institutions might hold off giving loans to let the rates rise higher, or because their cost of borrowing money is increasing. This would stifle growth in the economy, with companies forced to reduce investment. A vicious cycle, perhaps a version of a Minsky Moment.

There are other alternatives using the tax code. One option suggested by Martin Ford in *The Lights in the Tunnel* is modification of the payroll tax, a tax that discourages hiring people and encourages automaton since it makes the use of people more expensive. He suggests a reform of the tax system where we get away from taxing based on workers to reduce that disincentive for hiring. For example, he suggests a tax based on a company's gross margins, the difference between the price at which goods are sold and the cost of manufacturing such goods. To some degree, gross margin measures productivity, but can be impacted by many other factors, such as brand or product advantages that support high prices. It's less direct than the automation tax as proposed.

The automation tax might encourage companies to prefer productivity improvements achieved by using a combination of human and computer capabilities, as discussed at length previously. It's hard for an individual company to sacrifice for the good of all without feeling that competitors may not be so civic-minded—the automation tax is a feedback mechanism that makes such action rational.

16

Other Approaches to the Decline in Jobs

A tax isn't the only solution to the job shortage. There are other ways we can try to compensate for the effects of automation that have been proposed. I'm concerned that the approaches I outline in this chapter would take a long time to create results and or have unintended consequences, but, if they are more acceptable than a tax, they are better than ignoring our economic issues.

A shorter workweek

The forty-hour workweek (or so) has become almost a standard in the US. A shorter workweek means more people would be required to achieve the result of a longer workweek, in theory creating more jobs. Such an approach is usually accompanied by a compensating reduction in take-home pay for the worker by maintaining the hourly rate.

The approach requires that a company hire more employees to do the same work without compensating changes in the work to be done or the means to do it. This approach reduces disposable income per employee and reduces the median income, not a desirable outcome for the economy.

The new healthcare law, in an unintended consequence, is motivating this policy for another reason. The law requires employers over a certain size to provide healthcare insurance to *full-time* employees, and the insurance must have a minimum coverage level. Some companies with many low-wage employees, such as restaurant chains, hotels, and retailers, don't currently provide such coverage. Many are reacting to the law by trying to hire part-time employees (less than 32 hours per week) that wouldn't be covered by the law. This has the same effect as moving to a shorter workweek for those part-time employees.

The military

The defense budget can reduce expenditures on major weapon systems that are unlikely to be used in today's world and are very expensive, and apply those to hiring more soldiers, expanding their role beyond fighting wars. Increasing personnel in our military forces is a relatively easy way to create jobs without creating further bureaucracy. Typically, a soldier, unless he seeks a military career, will treat such a job as a transition to a job in the commercial sector. The military can make it part of their training to equip soldiers for that transition. My hope is that some of that training involves more instruction in computer literacy, which is clearly relevant to our defense forces today and is a skill transferable to the commercial sector.

The nature of a military force is such that part of their job is simply to maintain readiness. Part of that readiness can be training in dealing with emergency situations such as the impact of tropical storms, as previously noted. In the aftermath of tropical storm Sandy that struck the East Coast in 2012, the military helped in such areas as transporting fuel to areas that developed shortages. In a more extreme case, they might be required to maintain order and prevent looting. If the more extreme case was an emergency created by cyberwarfare, training to help with civilian problems such as electrical outages is even more appropriate.

I view this as a good option, but it is more likely to be used to maintain current levels of military forces rather than increase the number. Thus, it might not add new jobs.

Clean energy and distributed energy production

Jeremy Rifkin, in his 2011 *The Third Industrial Revolution*, among other points makes a persuasive case for the need to change our energy policy to avoid global warming, arguing that not doing so relatively quickly risks disastrous consequences. He joins many who have warned that we ignore this trend at our peril. Just before the 2012 election, New York mayor Michael Bloomberg, dealing with the damage of tropical storm Sandy, said, "Our climate is changing, and while the increase in extreme weather we have experienced in New York City and around the world may or may not be the result of it, the risk that it might be—given this week's devastation—should compel all elected leaders to take immediate action." A 2005 study by Webster, Holland, Curry, and Chang in *Science* reported that the number of 4 and 5 category storms has doubled since the 1970s, so we may find ourselves dealing with the penalties of global warming more quickly than we expect. Average temperature in the US in 2012 reached a new high.

Rifkin makes a good case for a distributed generation of energy, with individual homes and businesses generating energy locally, e.g., by solar panels on roofs or on the sides of large buildings, and contributing the excess back to the electrical network, all managed by Internet technology. Storage of energy to smooth peaks and valleys of generation and demand could occur through the use of hydrogen generation by any excess energy coupled with hydrogen-driven fuel cells. (Burning hydrogen creates water rather than carbon dioxide as a side result.) Obviously, such a major change would be slow and expensive. Rifkin cites policies in the European Union that encourage moving in this direction, with slower movement in the US.

In January 2013, the Los Angeles Department of Water and Power announced it would let customers sell back excess energy created on rooftops and parking lots. The program, with a focus on large multifamily dwellings, warehouses, school facilities, and parking lots, was described by the *Los Angeles Times* as "the largest urban rooftop program of its kind in the nation."

As a long-term solution to creating jobs, the energy conversion that Rifkin calls the Third Industrial Revolution can be a big contributor, with a large number of construction and other jobs created by the development. A 2008 study prepared by KEMA, an energy consulting firm, for the GridWise Alliance (a US coalition of IT companies, power and utility companies, academics, and venture capitalists promoting the "smart grid") found that $16 billion in government incentives supporting this goal would create $64 billion worth of projects and 280,000 direct jobs.

Another aspect of the distributed generation of power is that it, in effect, translates investments by individuals in such things as generating electricity, turning their roofs into lower energy costs for them and a source of income if they generate more than they use. With a smart grid in place, there is a mechanism for some families generating income with an investment that could be financed and/or subsidized, creating disposable income even without a job. Any investment is a risk, however, and we'd have to be cautious about assumptions of return on such an investment or we might have an equivalent to the mortgage bubble.

Rifkin's model of distributed energy generation is attractive, but it will take time. More immediate approaches to improve the economy such as the automation tax may improve the economic environment sufficiently to allow us the time to build a green energy economy.

17

Effective Use of Tax Revenues

The automation tax (or any other tax that generated additional revenue) could of course simply be used to reduce the federal deficit, and many would argue that that would help the economy in the long run (or even that it would help avoid economic collapse).

However, economists such as Paul Krugman in his 2012 book, *End This Depression Now!*, argue that increased government spending is necessary to get us out of the current slump. The automation tax won't necessarily reduce corporate profits in the long run if it is used to grow GDP and corporations sell more because of a revived economy. The suggestions in this section could be viewed as either a way to specifically allocate an automation tax or as a way for government expenditures to improve the economy whatever the source.

Promoting innovation

A nation's competitiveness depends upon a research base. In the US in particular, innovation has been a driving force maintaining our economic position in the world. Additional spending could be dedicated to research projects that help maintain innovation. One way would be increased support for university research, e.g., through increasing the budget of the National Science Foundation.

In addition, there are some government programs that have long been cost-effective in initiating new research outside academia. I've had good personal experience with both the Small Business Innovation Research (SBIR) program and the Defense Advanced Research Projects Agency (DARPA, part of the Defense Department). Both tend to offer relatively small grants compared to most government expenditures and programs, spreading the funding among many companies.

The SBIR program is coordinated by the US Small Business Administration. It was dropped for a while, but Congress voted to reauthorize the program and President Obama signed the bill into law in December 2011. Each year, Federal agencies with R&D budgets that exceed $100 million spent outside the agency are required to allocate 2.5% of their R&D budget to supporting research at small businesses. Each research contract for a given business is fairly small, but it can have a significant impact on the health and growth of a small business into a larger business. These are typically two-phase programs: the initial R&D grant is fairly small, with a second larger phase granted if the first phase is successful in meeting stated contract goals. The Small Business Technology Transfer Program is a similar program.

DARPA used to be called ARPA, and one well-known contribution was ARPAnet, the forerunner of the Internet. Without ARPA willingness to fund a national network connecting universities so that the advantages of the network became apparent, and so that universities were motivated to improve the technology, it would have taken much longer to develop the Internet, if it developed at all.

My involvement with DARPA was in a program where they held an annual meeting of researchers in speech recognition designed to discover which approaches were performing best. Previously, researchers would report in journals and conferences how well they were doing with different speech databases and applications, making it almost impossible to draw conclusions on which technology was actually working better. For example, recognizing ten digits is much simpler than recognizing a 1000-word vocabulary, and speech collected in a noisy environment is more difficult to recognize than speech collected in a quiet laboratory. Published papers reporting results on different tasks and in different environments made it hard for researchers to judge which technologies were better. DARPA supplied minimal funding to each participating research organization on the condition they participate in this annual meeting and run their technology on a speech database DARPA supplied, allowing an objective comparison of technologies. The result of this process was the conclusion that a method called Hidden Markov Models (HMMs) was far superior to other technologies. Although this program was in the 70's and 80's, HMMs are still the core technology used by most commercial speech recognition systems today.

Another program DARPA funded was called CALO (Cognitive Assistant that Learns and Organizes). Companies competed for the funds through proposals, and SRI International was a lead on the project. One result of SRI research was a spinoff company called Siri. The company Siri was purchased by Apple, and Siri is now well known as the voice personal assistant in the Apple iPhone. There are other spinoffs from the program that we are likely to hear more about. DARPA has several hundred personnel and about a $3 billion budget, a very small fraction of the Defense Department personnel and budget.

Government programs of this sort seem to get little publicity. They are successful because they build foundations rather than edifices. These are some of the best-run government programs we have, and increasing their budgets can improve US competitiveness at a very low cost. The very nature of these programs ensures that the funds would be spread widely across companies and research organizations, and the feedback principle would be in play, with selective rewards for those organizations that performed best in initial phases.

Social security

An aging population is an economic issue worldwide, both in terms of a social safety net and increased medical costs. The problem will only become more costly as improving medical technology allows living longer. In the US, supporting the cost of Social Security is a hotly debated long-term issue. And aging "baby boomers" are a known issue to the health of the program. The US Census Bureau projected in December 2012 that the percentage of the US population over 65 would rise to 22% by 2060 from the current 14%.

If a portion of any increased taxes is dedicated to the Social Security fund (to use the US case), it would help address the issue of an increasing number of people in the program. How does that reduce the economic problem? First, workers who are not yet at the retirement age, but close to it, may retire as planned, rather than keep working out of concern whether Social Security will be there for them. This makes more jobs available to others.

But, most importantly, seniors on Social Security are consumers. By maintaining or even raising their income, we create a growing class of consumers, growing as the population ages. I realize this is the opposite direction in which many analysts and politicians think Social Security spending should go, but cutting Social Security clearly would reduce disposable income for a growing group of consumers, and the spending would have to be made up somewhere to maintain the economy. If, despite all efforts, the number of jobs doesn't grow, seniors are one group who are not generally competing for jobs, but do form an important consumer group. It's not clear that reducing social security payments, a goal of some budget-balancing plans, has a long-term payoff; it may simply reduce the buying power of a growing consumer group, slow the economy, and reduce tax receipts. Social security may be a solution, rather than a problem.

Further, the current low interest rates, artificially induced by monetary policy, are reducing returns on individual retirement plans that adopt a conservative strategy, e.g., investing in interest-producing financial instruments such as treasury bills. The government is, in effect, stealing money that could be earned in retirement plans and indirectly reducing the assets in those plans upon retirement. This penalty hits the middle class the hardest, since those retirement funds can be the difference in maintaining their standard of living after retirement. Perhaps supporting social security, rather than shrinking it, would help pay back this theft from retirement plans.

Healthcare

The increase in healthcare costs and their impact on government programs such as Medicare is another issue that we have to face. More costly options for treating medical problems and an aging population are part of the challenge. Can we make this problem into a solution?

Perhaps. Healthcare is one of those industries where automation can reduce some costs and improve care, but it also requires many workers—doctors, nurses, pharmacists, dentists, and much more. Automation won't significantly reduce jobs in healthcare.

Many healthcare jobs require a significant investment in training, and are hard, time-consuming jobs. As a result, there appears to be a looming shortage of primary care physicians, as previously noted, in part because of the cost of a medical education. Government could provide education grants in some of these areas that might make it easier to generate qualified healthcare workers (and the jobs they represent). Some governments of course go much further and finance a national health system. Perhaps surprisingly, studies show that some of these government-managed systems appear to provide better healthcare at a lower cost than the US system.

Simply reducing payments to medical practitioners as a way of reducing Medicare costs will eventually result in less healthcare jobs as those challenging jobs don't compensate workers for the difficulty of their jobs and the training required to get them. And a shortage of healthcare workers will lead to a quality-of-care crisis. And, most likely, healthcare costs will eventually increase just to maintain quality of care. Already, many doctors are "going private," refusing to honor Medicare rates or charging an annual fee to each patient for the doctor being on call.

An article in the *Wall Street Journal* in 2010 forecast a "shortage of 150,000 doctors in 15 years," based on a report from the Association of American Medical Colleges. The trend is established—the number of medical school students entering family medicine in the US fell more than a quarter between 2002 and 2007. And importing new doctors from outside the US won't solve the problem; there is a shortage of residency positions in the US that creates a bottleneck, in that doctors must complete a US residency to practice in this country.

It might be effective to apply increased taxes in part to supporting Medicare as expenses rise. Spending more on healthcare creates more jobs and recognizes that rising healthcare costs are a result of need and of more options available for treatment when one is ill. Investing in healthcare serves a necessary societal goal. Again, I realize this is the opposite of most plans to reduce the US federal deficit, but there may be options to reduce it elsewhere. I also realize that there are some intrinsic problems that should be addressed in the US system beyond the financial issues.

Education

I discussed the crisis in education and the potential role of software in addressing the problem in an earlier chapter. Some money from increased taxes could go to education, supporting improved early reading education and adding computer literacy education. Both goals could be advanced by providing schools with tablet computers that could act as both textbooks and interactive teachers, allowing students to proceed at their own pace.

Preparing for and recovering from disasters

As an alternative to unemployment insurance, we can give unemployed people a job training for disaster recovery, such as the tropical storm Sandy that devastated the East Coast in October 2012, much like we maintain a trained military for possible wars. They can then work to help recover from such disasters more quickly when they occur. Once trained, they could remain in a reserve corps and called up in disasters. In fact, as ground wars become less likely, some soldiers could get this specialized training, as previously noted, providing justification for retaining military jobs as the apparent need declines.

Reduce personal tax rates

Any additional corporate tax could be coupled with a reduction in personal tax rates. More money in individual's pockets should lead to more spending, creating more jobs, more disposable income, and more sales for those corporations.

18

Building the Future

Making accurate predictions about technology has turned out over the years to be extremely difficult. Gordon Bell, a developer of early minicomputers (among other accomplishments), wrote an article on "The Folly of Prediction" in the book *Talking Back to the Machine: Computers and Human Aspiration*. He noted some errors of prediction made by people who should know better; e.g., a prediction in 1943 by Thomas Watson, the legendary chairman of IBM, that "there is a world market for maybe five computers." Bell's general conclusion is that it is difficult to project the future of technology due to unexpected developments. Nassim Nicholas Taleb, in *The Black Swan: The Impact of the Highly Improbable*, makes the same case.

The trends highlighted in this book are more of an observation than a prediction. The unexpected development has already occurred. It's the tightening of the human-computer connection through a coincidence of the rapid adoption of mobile devices and of language technologies breaking through a tipping point of utility. Software in modern technology enables the rapid acceleration of these trends. The impact on individuals and businesses is perhaps most evident in the explosive adoption of Internet-connected mobile devices. The impact on jobs of automation accelerated by software trends should be evident through careful examination of our economy, but seems masked by the assumption that the subprime crisis is the sole contributor to creating the current recession.

I've looked at these trends largely from two points of view: cultural evolution and the economic impact. In culture, I've noted the tightening of the human-computer connection, particularly through language. Technology is likely to be a constant companion—an always-available assistant—in the long term. Individuals and companies that recognize this trend can make the most of it. As a society, recognizing this trend gives us the option of optimizing our children's future through such things as reforming education. The coupling of human capabilities and computer intelligence is an innovation that has the potential to drive our economy to new levels.

This book has also examined other issues related to a software society. These include problems with the patent system and cyberwarfare. These issues can be addressed, but require action rather than passivity.

In economics, evidence indicates that, in the current recession, we have reached a point where automation is being used to avoid rehiring workers, creating a vicious downward economic cycle. We have the choice of changing the future by interceding with solutions that utilize the human-computer interface and motivate companies to create jobs. An automation tax can be used to provide that motivation, as well as to provide revenue for government programs that can be targeted at building jobs and consumption in other ways.

We can build the future rather than predict it. We can make the software society a new era of human growth.

Bibliography

Many of these references are explicitly cited in the body of this work. Those that aren't specifically cited provided insights and stimulation. All are recommended reading for readers who want to dig deeper into the themes addressed in this book.

Acemoglu, Daron and James Robinson, *Why Nations Fail: The Origins of Power, Prosperity, and Poverty*, Crown Business, 2012.

Amoroso, Edward G., *Cyber Attacks: Protecting National Infrastructure*, Butterworth-Heinemann, 2010.

Anderson, Chris, *Makers: The New Industrial Revolution*, Crown Business, 2012.

Arthur, W. Brian, "The Second Economy," McKinsey Quarterly, October 2011.

Arthur, W. Brian, *The Nature of Technology: What it is and how it evolves*, Free Press, 2009.

Ballmer, Steve, "Shareholder Letter," October 9, 2012, www.microsoft.com/investor/reports/ar12/shareholder-letter.

Bellman, Richard E., *Dynamic Programming*, Princeton University Press, 1957.

Bengtsson, Sara L., H. C. L., and Richard E. Passingham "Motivation to do Well Enhances Responses to Errors and Self-Monitoring," Cerebral Cortex **19**(4):7, 2009.

Berg Insight, *The Global Wireless M2M Market*, December 2009.

Berlinski, David, *The Advent of the Algorithm: The Idea That Rules the World*, Harcourt, Inc., 2000.

Bishop, Christopher M., *Pattern Recognition and Machine Learning*, Springer, 2007.

Bradley, Joseph, Jeff Loucks, James Macaulay, Richard Medcalf, and Lauren Buckalew, *BYOD: A Global Perspective*, Cisco, 2012.

Breiman, Leo, *Classification and Regression Trees*, Chapman & Hall/CRC, 1984.

Breiman, Leo, J. H. Friedman, R. A. Olshen, and C. J. Stone, *Classification and Regression Trees*, CRC Press, New York, 1999.

Brooks, Frederick P., *The Mythical Man-Month: Essays on Software Engineering*, Anniversary Edition, Addison-Wesley Professional, 1995.

Brynjolfsson, Erik and Andrew McAfee, *Race Against the Machine: How the Digital Revolution is Accelerating Innovation, Driving Productivity, and Irreversibly Transforming Employment and the Economy*, Digital Frontier Press, 2011.

Carr, Jeffrey, *Inside Cyber Warfare: Mapping the Cyber Underworld*, O'Reilly Media; 2nd edition, 2011.

Chang, Ha-Joon, *23 Things They Don't Tell You About Capitalism*, Bloomsbury Press, 2012.

Chou, Dorothy, *Transparency Report: Government requests on the rise*, Google Official Blog (http://googleblog.blogspot.com/2012/11/transparency-report-government-requests.html), November 13, 2012.

Citrix, "Workplace of the Future: A Global Market Research Report," September 2012.

Clarke, Richard A., Robert Knake, *Cyber War: The Next Threat to National Security and What to Do About It*, Ecco, 2012.

College Board, "Trends in College Pricing," 2012.

Congressional Budget Office, "Trends in the Distribution of Household Income Between 1979 and 2007," October 2011.

Costa, Rebecca, *The Watchman's Rattle: A Radical New Theory of Collapse*, Vanguard Press, 2012.

Datta, Saugato (editor), *Economics: Making Sense of the Modern Economy*, Wiley; 3rd edition, 2011.

Dawkins, Richard, *The Extended Phenotype: The Long Reach of the Gene*, Oxford University Press, 1999.

Dawkins, Richard, *The Selfish Gene*, Oxford University Press, 1976.

DeMarco, Tom, *Peopleware: Productive Projects and Teams*, Dorset House Publishing, 2010.

Denning, Peter J. (editor), *Talking Back to the Machine: Computers and Human Aspiration*, Springer, 1999.

Doyle, John C., Bruce A. Francis, and Allen R. Tannenbaum. *Feedback Control Theory*, Dover Publications, 2009.

Duhigg, Charles and Steve Lohr, "The Patent, Used as a Sword," The New York Times, Technology Section, October 7, 2012.

Evans, Peter C. and Marco Annunziata, "Industrial Internet: Pushing the Boundaries of Minds and Machines," GE. http://files.gereports.com/wp-content/uploads/2012/11/ge-industrial-internet-vision-paper.pdf, November 26, 2012.

Ford, Martin, *The Lights in the Tunnel: Automation, Accelerating Technology and the Economy of the Future*, CreateSpace Independent Publishing Platform, 2009.

Friedman, Thomas L., *The World is Flat: A Brief History of the Twenty First Century*, New York, Farrar, Strause and Giroux, 2005, 2006.

Giarratano, Joseph C., Gary D. Riley, *Expert Systems: Principles and Programming*, 4th edition, Course Technology, 2004.

Gratton, Lynda, *The Shift: The Future of Work Is Already Here*, Harper Collins, UK, 2011.

Guizzo, Erico and Evan Ackerman, "The rise of the Robot Worker," IEEE Spectrum, October 2012, pp. 34-41.

Hawkins, Jeff and Sandra Blakeslee, *On Intelligence*, Times Books, 2004.

Jarvis, Jeff, *Public Parts: How Sharing in the Digital Age Improves the Way We Work and Live*, Simon & Schuster, 2011.

KEMA, The US Smart Grid Revolution, KEMA's Perspectives for Job Creation, retrieved from www.kema.com/services/consulting/utility-future/job-report, December 23, 2008.

Keynes, John Maynard, *The General Theory of Employment, Interest and Money*, Palgrave Macmillan, London, 1936.

Keynes, John Maynard, *Treatise on Money and the General Theory of Employment, Interest and Money*, three volumes, 1927 to 1939.

Kirik, Cathie, "Accelerated Review of Green Technology Patent Applications," Inventors' Eye, (USPTO's bimonthly publication for the independent inventor community), June/July 2011.

Krugman, Paul, *End This Depression Now!*, W. W. Norton & Company, 2012.

Kurzweil, Ray, *The Singularity Is Near: When Humans Transcend Biology*, Penguin Books, 2006.

Lack, Simon, *The Hedge Fund Mirage*, Wiley, 2012.

Levy, Frank and Richard J. Murnane, *The New Division of Labor: How Computers Are Creating the Next Job Market*, Princeton University Press, 2004.

Lewis, F. L., *Applied Optimal Control and Estimation*, Prentice-Hall, 1992.

Lucas, Peter, Joe Ballay, and Mickey McManus, *Trillions: Thriving in the Emerging Information Ecology*, Wiley & Sons, 2012.

Martin, Robert C., *Clean Code: A Handbook of Agile Software Craftsmanship*, Prentice Hall, 2008.

Meisel, William (editor), *Speech in the User Interface: Lessons from Experience*, Trafford Press, 2010.

Meisel, William (editor), *VUI Visions: Expert Views on Effective Voice User Interface Design*, Trafford Press, 2006.

Meisel, William S., and Harold J. Payne, *Multi-Objective Control Theory*. Technical Report, Technology Service Corp., Santa Monica, CA, Jan-Dec., 1974.

Meisel, William, "Life on-the-Go: The Role of Speech Technology in Mobile Applications," chapter in *Advances in Speech Recognition: Mobile Environments, Call Centers and Clinics,* Neustein, Amy (Ed.), Springer, 2010.

Meisel, William, "The Personal Assistant Model: Unifying the Technology Experience," in *Mobile Speech and Advanced Natural Language Solutions*, Springer, 2013.

Meisel, William, "Ubiquitous Personal Assistants," Meisel-on-Mobile (blog), October 6, 2012.

Meisel, William, *Computer-Oriented Approaches to Pattern Recognition*, Academic Press, 1972.

Meisel, William, W. C. Liles, and M. D. Teener, *New Components and Subsystems for Digital Design*, Technology Service Corporation, 1975.

Meltzer, Allan, *Why Capitalism?*, Oxford University Press, 2012.

Minsky, Hyman P., *Can "It" Happen Again? Essays on Instability and Finance*, M. E. Sharpe, 1982.

Mishel, Lawrence, "The wedges between productivity and median compensation growth," Economic Policy Institute Briefing Paper #330, 12th edition of EPI's *The State of Working America*, 2012.

Mishel, Lawrence, and Josh Bivens, "Occupy Wall Streeters are Right About Skewed Economic Rewards in the United States," Economic Policy Institute Briefing Paper #331, 2011.

Mishel, Lawrence, Josh Bivens, Elise Gould, and Heidi Shierholz, *The State of Working America*, 12th edition. An Economic Policy Institute book. Ithaca, N.Y.: Cornell University Press, 2012.

Murphy, Kevin P., *Machine Learning: A Probabilistic Perspective*, The MIT Press, 2012.

Murray, Charles, "Why Capitalism Has an Image Problem," Wall Street Journal, July 27, 2012.

Myhrvold, Nathan, "I, Software," chapter in *Talking Back to the Machine*, 1999.

Neustein, A. and J. A. Markowitz (editors), *Mobile Speech and Advanced Natural Language Solutions*, Springer, 2013.

Neustein, Amy (editor), *Advances in Speech Recognition: Mobile Environments, Call Centers and Clinics*, Springer, 2010.

O'Regan, Gerard, *A Brief History of Computing*, Springer, 2008.

Payne, H. J., M. D. Teener, and William Meisel, "Ramp Control to Relieve Freeway Congestion Due to Traffic Disturbances," Technology Service Corporation, Report No. 034, April 1972.

Posner, Richard, "Do patent and copyright law restrict competition and creativity excessively?", Becker-Posner Blog (www.becker-posner-blog.com), September 30, 2012.

Rathbun, A., and J. West, "From Kindergarten Through Third Grade: Children's Beginning School Experiences" (NCES 2004-2007). National Center for Education Statistics, Institute of Education Sciences, US Department of Education, Washington, DC, 2004.

Ravitch, Diane, *The Death and Life of the Great American School System: How Testing and Choice Are Undermining Education*, Basic Books, 2011.

Reich, Robert B., *Beyond Outrage: What has gone wrong with our economy and our democracy, and how to fix them*, Knopf, 2012.

Rifkin, Jeremy, *The End of Work*, J. P. Tarcher, 1995.

Rifkin, Jeremy, *The Third Industrial Revolution: How Lateral Power Is Transforming Energy, the Economy, and the World*, Palgrave Macmillan, 2011.

Romer, P. M. "Why, Indeed, in America? Theory, History, and the Origins of Modern Economic Growth," American Economic Review **86**(2), 1996.

Romer, Paul, "When Should We Use Intellectual Property Rights?", American Economic Review, 92(2): 213-216, 2002.

Romer, Paul M., "Endogenous Technological Change," *The Journal of Political Economy*, Vol. 98, No. 5, Part 2: The Problem of Development: A Conference of the Institute for the Study of Free Enterprise Systems, pp. S71-S102, Oct., 1990.

Roubini, Nouriel and Stephen Mihm, *Crisis Economics: A Crash Course in the Future of Finance*, Penguin, 2011.

Sataline, Suzanne and Shirley S. Wang, "Medical Schools Can't Keep Up," Wall Street Journal, April 12, 2010.

Schumpeter, J. A., *Capitalism, Socialism, and Democracy*, 3rd edition (originally published in 1950), introduction by Thomas K. McCraw, Harper Perennial Modern Classics, 2008.

Schumpeter, J. A., *The theory of economic development: an inquiry into profits, capital, credit, interest, and the business cycle*, translated from the German by Redvers Opie 1961.

Shapiro, Robert J., *Futurecast: How superpowers, populations, and globalization will change the way you live and work*, New York, St. Martin's Press, 2008.

Shore, Genevieve, "Why disruption is good for business," BBC web article, www.bbc.co.uk/news/business-20042737, October 25, 2012.

Sirkin, Harold L., Justin Rose, and Michael Zinser, *The US Manufacturing Renaissance: How Shifting Global Economics Are Creating an American Comeback*, Knowledge@Wharton, 2012.

Sowell, Thomas, *Basic Economics: A Common Sense Guide to the Economy, 4th edition*, Perseus Books Group, Dec. 28, 2010.

Spence, Michael, *The Next Convergence: The Future of Economic Growth in a Multispeed World*, Farrar, Straus and Giroux, 2011.

Steiner, Christopher, "Automate This: How Algorithms Came to Rule Our World," Portfolio/Penguin, 2012.

Taleb, Nassim Nicholas, *The Black Swan: The Impact of the Highly Improbable* (Second Edition), Random House, 2010.

Thurber, Kenneth J., *Do NOT Invent Buggy Whips: Create, Reinvent, Position, Disrupt*, Digital Systems Press, 2012.

Tucker, Robert C. (editor), *The Marx-Engels Reader*, 2nd edition, W. W. Norton & Company, 1978.

Turkle, Sherry, *Alone Together: Why We Expect More from Technology and Less from Each Other*, Basic Books, 2012.

Vinge, Vernor, "Vernor Vinge on the Singularity," Web: http:// mindstalk. net/ vinge/ vinge-sing.html, 1993.

Walker, David, *Comeback America: Turning the Country Around and Restoring Fiscal Responsibility*, Random House, 2009.

Warsh, David, *Knowledge and the Wealth of Nations: A Story of Economic Discovery*, W. W. Norton & Company (2007).

Webster, Peter, G. Holland, J. Curry, & H. Chang, "Changes in Tropical Cyclone Number, Duration, and Intensity in Warming Environment," *Science,* 309 (5742), 1844-1846, 2005.

Whiten, Andrew, et al., *Current Biology*, 17, 1038-1043, 2007.

Whittaker, James A., Jason Arbon, and Jeff Carollo, *How Google Tests Software*, Addison-Wesley Professional, 2012.

Wildavsky, Ben (author, editor), Andrew Kelly and Kevin Carey (editors), *Reinventing Higher Education: The Promise of Innovation*, Harvard Education Press, 2011.

Wolff and Sherman, *Rodent Societies: An Ecological and Evolutionary Perspective*, 2007.

Yudkowsky, Moshe, *The Pebble and the Avalanche: How Taking Things Apart Creates Revolutions*, Berrett-Koehler Publishers, 2005.

Index

About the Author

William S. ("Bill") Meisel's experience combines a strong academic, technical, and business background. With a B.S. degree in Engineering from Caltech and a Ph.D. in Electrical Engineering from the University of Southern California, he began his career as a professor of Electrical Engineering and Computer Science at USC. He has published over 70 papers and several books, including a technical book, *Computer-Oriented Approaches to Pattern Recognition*. He was the editor and contributor to *New Components and Subsystems for Digital Design*, published in 1975; *VUI Visions: Expert Views on Effective Voice User Interface Design*, published in 2006; and *Speech in the User Interface: Lessons from Experience*, published in 2010.

After working in university research, he formed and managed the Computer Science Division of an aerospace company for ten years, applying pattern recognition technology to a number of application areas, including target detection and intelligence data analysis. He then founded a venture-capital-backed speech recognition company and ran it for ten years. He has been issued six patents in various areas of technology. Currently, as an industry analyst, he publishes a paid-subscription, no-ads industry newsletter on commercial developments in speech technology and natural language interpretation. As Executive Director of the Applied Voice Input Output Society (AVIOS), a non-profit industry organization, he creates the program for the annual *Mobile Voice Conference*. He also consults for several technology companies.

www.ingramcontent.com/pod-product-compliance
Lightning Source LLC
Chambersburg PA
CBHW051227050326
40689CB00007B/827